The Global Recruiter's Guide to the U.S. IT Industry

Strategies, Skills and
Success for Talent Partners
Worldwide

AUTHOR: JAY BARACH

The Global Recruiter's Guide to the U.S. IT Industry:
Strategies, Skills, and Success for Talent Partners Worldwide

Copyright © Jay Barach (2025)

ISBN Paperback: 979-8-9991927-0-7
ISBN eBook: 979-8-9991927-1-4

DEDICATION

To my father, Mr. Shailendra

You always believed in me and encouraged me to embrace the philosophy that no one in this world is perfect yet we must strive for perfection. Because while small things make perfection, perfection is not a small thing. No words can express the depth of my gratitude for your unwavering support and the countless sacrifices you've made for me.

Even if I don't always express it, I love you and hold the deepest respect for everything you are and all that you've done.

And to every aspiring recruiter who dared to dream beyond the transaction,

To every talent acquisition specialist who chose partnership over placement,

To every hiring manager who believed in collaboration, not control,

And to every candidate who trusted the process and never lost hope,

This book is for you.

Table of Contents

PREFACE

In today's interconnected world, recruitment is no longer bound by geography. The rise of remote work, global talent pools, and digital hiring platforms has reshaped how companies attract, evaluate, and retain talent especially in the fast-paced world of information technology (IT). At the same time, professionals entering the field of talent acquisition are expected to navigate complex client expectations, evolving technologies, and cultural nuances with both speed and accuracy.

This book is designed as a practical guide for aspiring and early-career recruiters and talent acquisition specialists across the globe, including those working in or with the U.S. IT industry. Whether you're based in New York or New Delhi, Chicago or Cebu, the foundational principles of effective recruiting remain the same: understand the role, know your domain, build trust, and deliver consistently.

Having worked at the intersection of IT, project management, and recruitment, I've witnessed firsthand how many professionals, especially those transitioning into U.S.-focused recruitment struggle to bridge the gap between technical complexity and human connection. This book exists to close that gap. It is built on lived experience, real-world challenges, and the insights I've gained by serving clients, guiding candidates, and mentoring recruiters.

You'll find examples, industry-specific breakdowns, self-assessments, and strategic advice, each crafted to not only teach, but to empower you to think like a talent partner, not just a resume pusher.

Whether you're sourcing your first candidate, managing a VMS client, or planning to build a career in strategic talent acquisition, this book will help you develop the skills, mindset, and ethical clarity needed to succeed.

Welcome to a journey that's about more than placing candidates. It's about creating value for your clients, your candidates, and your own professional growth.

—Jay Barach

ACKNOWLEDGEMENTS

Writing this book has been a journey of deep reflection, research, and purpose. I owe immense gratitude to the many individuals who inspired, supported, and guided me throughout this process. Their encouragement, expertise, and friendship shaped this project into what it is today. Whether through conversations, mentorship, collaboration, or belief in the vision of this work, each of you has played an invaluable role.

With sincere appreciation, I acknowledge:

Beth. V
President / Chief Executive Officer,
Systems Staffing Group, Inc.

David. V
Chief Financial Officer,
Systems Staffing Group, Inc.

Karen. G
Account Executive/Recruiter,
Systems Staffing Group, Inc.

Robert. S
Sr. Talent Acquisition Specialist,
GSTi

Jainish. P
Managing Partner,
BMS1 Solution

Aayushi. B
Director of Finance and Human Resources,
BMS1 Solution

Kashyap S.
Director of Marketing and Business
Development,
BMS1 Solution

Late Darshil. S
Managing Partner,
Rise RPO

Deepak Sevkani
Lead Recruiter,
Rise RPO

Viveka. K
Vice President,
CEC

Pushpender. C
Chief Executive Officer,
CEC

Ravi. S
Recruitment Lead,
Motion Recruitment

Jayant. P
Owner / Proprietor,
Shubh Labh Enterprise

Jaipal. G
Head IT,
Tata Power Company

Shailesh. R
Project Manager,
Tata Consultancy Services

Rajesh. K
Sr. Network Engineer,
Tata Power Company

Sanjeev. J
Scientist SF,
Space Applications Centre, ISRO

Vikas. S
Founder,
Networkers Home

Harsh B.
Senior Manager,
GEA Group

Mitul B.
Entrepreneur,
Mitu Fashion House

INTRODUCTION

The world of recruitment has undergone and is still undergoing a remarkable transformation. Before, recruitment was limited to the local talent pools, but those days are gone. Today, recruitment is no longer limited; it is now a global market. We are now in a world where continents connect, language barriers are no longer a problem in the face of glaring talent, and talent goes across different countries' borders.

The increasing demand for specialized skills, technology developments, the need for better incomes, and the rise in remote work drive this change in the recruitment landscape. As a result, most companies are no longer restricted to hiring talents from their local vicinity. Instead, they can now tap into a global pool of increasingly talented people, leveraging the best skills and expertise from everywhere worldwide. The implications of these transformations are wide-ranging. For example, it has served as a significant source of income and created opportunities for job seekers and unemployed people, who can now explore job opportunities beyond their geographical locations. At the same time, it has also raised the bar for companies that must compete with global peers to attract and retain top talent, thereby increasing revenue for the company and satisfying both workers and clients. It has also led to a rise in the country's economies and appreciation of their different currencies.

In this new world of global recruitment, language barriers and cultural differences are no longer tangible obstacles. Instead, they are now held with a tinge of pride as a company with a large workforce from different cultures and backgrounds, which is considered more developed as it comprises different cultural knowledge coming together to create magic. More opportunities are presented to companies, allowing them to tap into other perspectives, ideas, and experiences. As these individuals come from different locations, they bring new skills, knowledge and insights that could help the companies innovate, adapt better, and thrive in an increasingly complex work environment and the interconnected world.

As we go through this continuously changing and brave new world of global recruitment, it is essential to note and understand the trends, challenges and opportunities shaping it. From this, a pressing question arises:

"How can we, as professionals in talent acquisition, understand and adapt to the ever-changing demands of the global recruitment market? And how can we stay ahead of the curve?"

In forthcoming chapters, we will delve into the primary factors of global recruitment, explore the strategies and methods these companies use to attract and retain top talents and evaluate the effects of this transformation on job seekers, companies, and the future of the workforce itself.

With its dynamic growth and constant need for specialized roles, the US IT industry presents a unique challenge. To meet this demand, talent acquisition professionals from different areas, including Asia, Europe, Africa, Oceania, and even the Americas, must come together to share their expertise and knowledge and learn from each other.

This book is your step-by-step companion, guiding you through the complex state of the US IT recruitment space. It is a must-read if you are interested in or ready to enter the US IT recruitment space or scale your impact within it. Whether you are a complete beginner, a career changer, or an experienced recruiter looking to level up and become a true talent advisor, this book has got you covered.

So, what sets this book apart?

Recruitment strategies are often reduced to buzzwords and other trendy concepts, but this book takes a different approach. We are going beyond the buzzwords and delving into the real-world applications of recruitment strategies. You will gain insight into the following:

- The US hiring manager's mindset: To succeed in recruitment, you must understand what controls hiring managers' decisions. Some questions that you should ask yourself include: What are their challenges? What do they value most in a candidate? How can you be the Candidate they want to hire? By gaining insight into the hiring manager's mindset, you can understand and match your approach to meet their unique needs and preferences, increasing your chances of being hired.

How technologies like ATS and AI affect your job: Technology is transforming the recruitment landscape. From Applicant Tracking Systems (ATS) to Artificial

Intelligence (AI) and other automation tools, there are many ways to facilitate your recruitment process and stay ahead of the competition. We will explore the latest technologies and strategies for using them to achieve your recruitment goals.

- What makes Boolean Search powerful: Boolean Search is a powerful tool for recruiters because it allows them to refine search results, increase search accuracy, save time, find hidden talent, combine multiple criteria, search across multiple platforms, etc.

Why Domain knowledge is critical: Before doing anything, it is always essential to have domain knowledge about it. Domain knowledge is necessary during recruitment, even for non-technical recruiters. By developing your expertise in a specific industry or niche, you can better understand the needs of your clients and candidates and provide more effective solutions. We will also explore the importance of domain knowledge and provide practical tips for developing your expertise.

- How to stand out in a Saturated recruitment market: In a competitive recruitment market, getting lost in the crowd is always very easy, especially when everyone is doing something similar. To succeed, you must differentiate yourself and provide unique value to your clients and candidates. We would explore strategies for standing out in a saturated market, from developing a personal brand to creating innovative solutions that would meet the evolving needs of the recruitment landscape and become a true talent advisor.

Others include the future of recruitment, the art of storytelling, and the importance of data-driven decision-making in recruitment.

To ensure that you get the most out of this book, we have carefully designed each chapter with the following elements:

1. Clear Objectives: At the beginning of each chapter, we have outlined clear objectives highlighting what you should expect to learn. These objectives serve as a map to guide you through the chapter's content and understand how it applies to your role in recruiting IT professionals for US clients.

2. Real-world context from Cross-Continental hiring scenarios: To make the learning experience more relatable and engaging, we have incorporated real-world context from cross-continental hiring scenarios. These scenarios will demonstrate how recruitment strategies and best practices are applied in different regions, cultures, and industries worldwide and how these are utilized in recruitment now.

3. Practical Diagrams: They are a way of visualizing complex concepts. Complex concepts can be challenging to understand, especially for those new to recruitment or talent acquisition. To address this, we have also included practical diagrams like:
   - Organization charts (org charts) to illustrate company structures and hierarchies
   - Search logic trees to demonstrate how to construct effective search queries
   - Visa flowcharts to outline the steps involved in navigating US visa processes

   These diagrams visually represent complex concepts, making it easier for you to understand and apply them in your daily role.

4. Self-assessment tests to check your growth: The goal of learning is to grow, and of course, one of the ways of measuring your growth is through assessments. We have included self-assessment tests throughout the book to help you evaluate your understanding and retention of the material. These tests enable you to:
   - Identify areas where you need more practice or review.
   - Track your progress and growth as you work through the chapters to know how well you understand them.
   - Reinforce your learning and build confidence in your abilities.

5. Templates and toolkits to use in your everyday role: To support your work in recruiting IT professionals for US clients, we have also provided templates and toolkits for your everyday role. Some of these resources include:
   - Sample job descriptions and job postings
   - Interview guides and assessment templates
   - Visa application checklists and timelines

- Recruitment metrics and reporting templates

These practical resources will help you facilitate recruitment processes, improve efficiency, and get better results.

WHY THIS BOOK EXISTS

This book addresses the growing need for skilled recruiters and talent acquisition professionals who can easily go through all the difficulties and processes of recruiting IT professionals for US clients. It serves as a way to bridge the knowledge gap. Despite the increasing demand for talent, many recruiters and acquisition professionals struggled to find and attract the right candidates. This is often due to a lack of understanding of the US job market, the process of getting a visa, and the challenges of recruiting IT professionals. It also serves as a way of providing practical guidance by using real-life examples that are very relatable and expert insights. It also empowers recruiters and talent acquisition professionals and helps them improve their skills, increase their potential to earn higher and contribute to the company's growth and success. It can also address global talent shortage as it would teach IT recruiters ways of discovering and identifying hidden talents from everywhere and how to develop effective recruitment methods.

WHO THIS BOOK IS FOR

This book has been written to be a valuable resource for anyone involved or interested in recruiting IT professionals for US clients. It doesn't matter if you are starting your career or hungry for more knowledge. This book is for the following set of people:

1. Career Switchers: Are you looking to transition into a new recruitment or talent acquisition career? This book is an ideal source of knowledge for career switchers as it provides much insightful knowledge and a well-explained introduction to the world of IT recruitment to enable you to learn the fundamentals of recruitment, discover how to leverage skills such as problem-solving skills, communication skills, project management skills etc., to succeed in recruitment.
2. Freshers (New Recruits): Are you new to recruitment and talent acquisition? This book is designed to give people with zero knowledge

some ground. It teaches everything about IT recruitment with a very comprehensive guide. It enables recruits to learn the basics of recruitment, how to manage client relationships, understand technical requirements, manage stakeholders' expectations, and meet deadlines. It also enables them to gain confidence in their abilities to succeed in recruitment using tips and expert advice.

3. Agency Recruiters: Are you an agency recruiter looking to improve your skills and knowledge in IT recruitment? Then, this book is for you as well! It provides a guide to recruiting IT professionals for US clients, including Strategies to source talents and how to leverage social media, job boards, and professional networks. It also discusses techniques to build effective and strong relationships with clients, manage expectations, and provide exceptional customer service. Finally, it explains how to stay ahead of the competition and adapt to the continuous changes in the global market.

4. Internal Talent Acquisition (TA) Teams: Are you part of a company's internal talent acquisition team? Are you looking to expand your horizons? This book is for you, too. It will guide you to an effective recruitment marketing campaign, insights, and many adaptation techniques.

5. Human Resource Professionals and Managers, etc.

WHAT MAKES OUR RECRUITMENT DIFFERENT?

Recruiting of IT professionals might present some challenges and opportunities. Some factors which lead to the distinct nature of US recruitment include:

- Complex Visa Processes: The US visa process can be quite time-consuming, requiring recruiters to go through various visa categories, eligibility requirements and application procedures.

High demand for specialized skills: The US IT industry requires some specialized skills, which prompts recruiters to seek top talents in areas like cloud computing, artificial intelligence, and cyber security.

- Competitive Job Market: The US job market is highly competitive, with multiple companies vying for top talent. Recruiters must develop effective strategies to attract and retain candidates to prevent them from switching to the nearest competitors or the company's rival.

Diverse Talent Pool: The US talent pool has candidates from various cultural, educational, and professional backgrounds. Recruiters must adapt their approaches to fit into this pool.

- Strict Labor Laws and Regulations: The US labour laws and regulations, such as the Fair Labor Standards Act (FLSA) and the Immigration and Nationality Act (INA), govern the recruitment and employment process. Recruiters must obey these rules and regulations no matter what.

Advanced recruitment technologies: The US recruitment landscape is characterized by advanced technologies, such as Applicant Tracking Systems (ATS), Artificial Intelligence (AI), and social media platforms. Recruiters must use these technologies to facilitate their processes and ensure they stay competitive.

- High Expectations for Candidate Experience: US candidates expect a seamless and personalized recruitment experience. Recruiters must prioritize candidate experience, ensuring timely communication, transparency and respect throughout the hiring process.
- Global Competition for Talent: The US recruitment market is not isolated; it competes with other global markets for top talent. Recruiters must develop strategies to attract and retain international candidates.

Evolving nature of work: The US workforce is undergoing significant changes, which are controlled by factors like remote work, the gig economy, and shifting employee expectations. Recruiters must adapt to these changes and develop flexible and innovative recruitment strategies.

Cultural norms: Cultural differences can significantly impact the workplace. For instance, communication styles, work habits, and expectations vary across cultures. Understanding these differences is essential for effective team management and communication.

Business Practices: Business practices in the US are influenced by the country's capitalist economy and emphasis on innovation. Companies prioritize competitiveness, efficiency, and customer satisfaction. In this context, it means finding candidates with the right skills and the ability to adapt and survive in a continuously developing environment.

(HTTPS://workmotion.com)

THE CAREER POTENTIAL OF TALENT ACQUISITION IN THE GLOBAL TECH WORLD

Talent acquisition has vast and exciting career potential in the global tech world. With the rise of remote work, companies are now looking for talent beyond geographical boundaries, making it an ideal time to pursue a career in talent acquisition.

What is happening in the Global Tech World?

- Global talent acquisition is on the rise: Companies are no longer limited to hiring talent from their local area. They are now looking for the best candidates from all over the world.
- New talent hubs are emerging: Countries like Uruguay and Poland are becoming popular destinations for companies looking for skilled professionals.

AI and automation are changing the game. These technologies are helping talent acquisition professionals streamline their processes, improve the candidate experience, and make data-driven decisions.

Diversity and inclusion are top priorities: Companies are now more focused than ever on building diverse and inclusive teams. This means that talent acquisition professionals are crucial in developing strategies that attract and retain diverse talent.

What Kind of Roles Can You Expect?

Talent Acquisition Specialist: You'll find, attract, and hire the best talent.

Global Talent Acquisition Manager: You will oversee global talent acquisition strategies and ensure they align with business objectives.

- Recruitment Analyst: You will analyze data to inform recruitment strategies and identify areas for improvement.

What Skills Do You Need?

- A global mindset: You should be able to understand and adapt to different cultures and talent pools.

- Data analysis skills: You should be able to collect, analyze, and interpret data to inform talent acquisition strategies.
- Excellent communication skills: You should be able to build strong relationships with stakeholders, candidates, and hiring managers.

Technical skills: You should be proficient in recruitment software, AI-powered tools, and data analytics platforms.

This book provides the insights, strategies, and tools you need to succeed in the rapidly evolving recruitment landscape. Whether you are a seasoned recruiter or just starting out, this book will help you stay ahead of the curve and achieve your goals. Throughout this book, we will share real-life examples, practical diagrams, and templates to support your learning. You will also be able to assess your knowledge and skills through self-assessment tests.

So, are you ready to embark on this journey and become a global recruitment expert? Let's get started!

CHAPTER ONE

BECOMING A GLOBAL TALENT PARTNER

The objectives of this learning:

- Understand the evolution from transactional recruiting to strategic talent partnership
- Learn why global recruiters are critical to the US IT talent ecosystem
- Identify the key differences between a recruiter and a talent acquisition specialist
- Recognize essential skills and traits of a successful global recruiter
- Discover how belief in the product (your Candidate and the opportunity) influences success

THE EVOLUTION OF RECRUITING FROM TRANSACTIONAL TO STRATEGIC

What was once a transactional process focused on filling job openings and granting employment has evolved into a strategic partnership that drives business growth and success.

Transactional Recruiting:

Transactional recruiting refers to a traditional approach focusing on filling job openings faster and more efficiently. This approach is often characterized by:

1. It's focused on filling job openings: The primary goal is to fill job openings as quickly as possible, and most times, without considering the long-term implications of the hire.
2. Emphasis on speed and efficiency: Transactional recruiters mainly prioritize speed and efficiency, using mass emailing, job board posting, and resume screening.
3. Limited relationship building: Transactional recruiters focus on the immediate need rather than building long-term relationships with candidates, hiring managers, or stakeholders. This leads to the recruiters not knowing what goes on with the candidates and focusing on the income alone.

4. Reactive approach: Transactional recruiters often react to job openings rather than consider future talent needs. They only consider immediate employment.

Tactics Used in Transactional Recruiting include:

1. Job board posting: Posting job ads on popular job boards to attract a large pool of candidates.
2. Resume screening: Quickly scanning resumes to identify potential candidates.
3. Mass emailing: Sending general emails to large groups of candidates for publicity and getting feedback to identify potential candidates.
4. Phone screening: Conducting brief phone interviews to assess the different Candidate's qualifications.

Limitations of Transactional Recruiting;

1. Poor candidate experience: Transactional recruiting can lead to a poor candidate experience, as candidates may feel like they are just a number rather than a valued individual. In other cases, the company keeps recruiting anybody without knowing how experienced they are.
2. Low-quality hires: Focusing solely on speed and efficiency can result in low-quality hires, as recruiters may overlook critical qualifications or cultural fit and only care about how fast they can get results. This can lead to lazy or rushed jobs, which, in most cases, would be below expectations.
3. Lack of strategic value: Transactional recruiting often fails to provide strategic value to the organization, as recruiters are not considering the long-term implications of their hires. They are only focused on short-term goals and ruses results.

When to Use Transactional Recruiting:

1. High volume hiring: Transactional recruiting can be effective for hiring many people in cases when speed and efficiency are crucial.
2. Entry-level positions: Transactional recruiting may be suitable for entry-level positions, focusing on finding candidates with basic qualifications rather than highly qualified or experienced candidates.

However, for most organizations, a more strategic approach to recruiting is necessary to attract and retain top talent.

Strategic Recruiting:

Strategic recruiting is a proactive and thoughtful approach to talent acquisition. It involves aligning recruitment efforts with the organization's business strategy, goals, and objectives. It is often characterized by:

1. Long-term focus: Strategic recruiting involves thinking ahead, anticipating future talent needs, and developing a plan to meet those needs.
2. Alignment with business objectives: Recruitment efforts are always aligned with the business strategy, goals, and objectives, ensuring a smooth run.
3. Proactive approach: Strategic recruiters are proactive. They anticipate talent needs and develop a pool of qualified candidates.
4. Relationship building: Strategic recruiters build long-term relationships with candidates, hiring managers, and stakeholders.
5. Data-driven decision-making: Strategic recruiters use data and analytics to inform recruitment strategies and control business outcomes.

Tactics Used in Strategic Recruiting include:

1. Talent pipelining: Building pipelines of qualified candidates to meet future talent needs
2. Employer branding: Developing and promoting the employer's brand to attract top talents and increase the workforce.
3. Candidate engagement: Building relationships with candidates through social media, events, and other channels.
4. Market research: Conducting market research to understand talent trends, competitor activity, and industry developments in the global market.
5. Analytics and reporting: Using data and analytics to measure recruitment effectiveness and control business outcomes.

The benefits of Strategic Recruiting are:

1. Improved quality of Hire: Strategic recruiting leads to better quality hires, as recruiters can attract and select top talent instead of wasting so many resources on quantity rather than quality.
2. Reduced time to hire: Strategic recruiting reduces hiring time, as recruiters already have a pool of qualified candidates and can move quickly. This helps them save time without going through the stress of interviewing or screening candidates to pick out the qualified ones, as they are all equally skilled.
3. Cost savings: Strategic recruiting can lead to cost savings, as recruiters can reduce reliance on external agencies and job boards to attract candidates.
4. Enhanced employer brand: Strategic recruiting enhances the employer brand, as recruiters can promote the organization's values, mission, and culture rather than having many candidates who are focused on their personal gains.
5. Better alignment with business objectives: Strategic recruiting ensures that recruitment efforts align with the organization's overall business goals and objectives, as the recruiter candidates are qualified and sometimes experienced enough to know what the organization needs to grow.

When to Use Strategic Recruiting:

1. Critical roles: Strategic recruiting is essential for critical roles, where the quality of hire can significantly impact business outcomes.
2. Competitive markets: Strategic recruiting is necessary in competitive markets, where organizations need to differentiate themselves to attract top talent. Most candidates are interested in companies that stand out and may not want to be with companies that keep doing the same innovations with other companies over and over again.
3. Long-term growth: Strategic recruiting is essential for organizations focused on long-term development, where talent acquisition is critical to achieving business objectives.

The Evolution of Recruitment from Transactional to Strategic.

To understand this evolution, we need to take a step back. Traditionally, recruiters were tasked with finding and hiring candidates. Their role was largely transactional, focused on filling job openings rather than understanding the client's broader business needs. However, as the job market became increasingly complex and the demand for skilled talent skyrocketed, recruiters needed to adapt. They had to develop a deeper understanding of their clients' businesses, anticipate market trends, and deliver more strategic value. This is why the evolution occurred, as switching from transactional to strategic was essential.

WHY GLOBAL RECRUITERS ARE CRITICAL TO THE US IT TALENT ECOSYSTEM

Here are some reasons why global recruiters are critical to the US IT talent ecosystem:

1. Access to Global Talent Pool:

- Diversity of skills: Global recruiters can tap into the diverse pool of skills and expertise from around the world, which would help bring unique perspectives and experiences to US IT companies.

Scalability: With a global talent pool, recruiters can measure their recruitment efforts more easily. This will help them meet the growing demand for IT talent in the US, as they can easily tell when they are lagging.

2. Addressing IT Talent Shortage:

- Shortage of skilled workers: The US IT industry has a shortage of skilled workers, particularly in areas like artificial intelligence, cybersecurity, and data science. Global recruiters can help bridge this gap as recruiters come from different parts of the world, some of which are filled with much knowledge.

Talent competition: IT talent in the US is very fierce. Global recruiters can help US companies stay competitive by accessing talent from other regions or countries, as talent in the US might begin to have a hard time choosing where they want to be and might just reject all.

3. Driving Innovation and Growth

- Innovation through diversity: Global recruiters can bring together diverse teams of IT professionals worldwide, bringing about innovation and creativity in the US IT industry.

Access to new markets: By recruiting global talent, US IT companies can access new markets, customers from different regions, and revenue streams, increasing business growth and expansion.

4. Enhancing Employer Brand

- Global employer brand: Global recruiters can help US IT companies develop a strong global employer brand, attracting top talent worldwide. This is because the employer brand is now a popular name. For example, a talented model would rather work for popular clothing brands like Gucci, which has brands worldwide and already has a good name, instead of working for a nameless brand or a brand that isn't global yet because of the attached benefits.

Diversity and inclusion: By recruiting global talent, US IT companies can demonstrate their commitment to diversity and inclusion, enhancing their reputation and attractiveness to top talent. They know that no matter what region they are from, they will feel included and seen.

5. Overcoming Recruitment Challenges

- Speed and efficiency: Global recruiters can help US IT companies overcome recruitment challenges by providing speed and efficiency in hiring.
- Cultural and language barriers: Global recruiters can navigate cultural and language barriers, ensuring that US IT companies can access the best talent worldwide.

KEY DIFFERENCES BETWEEN A RECRUITER AND A TALENT ACQUISITION SPECIALIST

There is a common misconception that a recruiter and a talent acquisition specialist are the same, but they are definitely not the same. Though both are involved in attracting and hiring top talents, there are differences between the two, and that is what we will discuss here.

The key differences between a recruiter and a talent acquisition specialist include the following:

KEY DIFFERENCES BETWEEN A RECRUITER AND A TALENT ACQUISITION SPECIALIST

There is a common misconception that a recruiter and a talent acquisition specialist are the same, but they are definitely not the same. Though both are involved in attracting and hiring top talents, there are differences between the two, which is what we will discuss here.

The key differences between a recruiter and a talent acquisition specialist include the following:

RECRUITER	TALENT ACQUISITION SPECIALIST
1 Focuses on filling job openings.	1 Strategically partners with hiring teams.
2 Often works reactively.	2 Often works proactively. Building long-term talent pipelines.
3 Relies heavily on job boards and basic screening.	3 Uses data, branding and Candidate experience as tools.
4 Transaction oriented.	4 Relationship oriented.
5 Limited business knowledge.	5 Business knowledge.

Now, we will be elaborating more on these differences.

RECRUITERS.

1. Focuses on filling job openings: A recruiter's primary focus is filling open positions quickly. His main priority is speed and employing many people instead of recruiting qualified and experienced people. This only fits immediate needs and can never be used for the company's long-term goals.

2. Often works reactively: Recruiters often work reactively; they react quickly to job openings or vacancies as they arise instead of proactively anticipating future talent needs. This can lead to last-minute hiring because the previously hired couldn't get the job done, there is an inefficient use of resources, and there is a poor fit between the hired Candidate and the company's goals, etc. A reactive approach can

hinder long-term success and make adapting to changing business needs challenging.

3. Relies heavily on job boards and basic screening: Job boards are when job openings are posted on a popular board to attract candidates. This attracts many applications from different people, both qualified and unqualified. Basic screening involves reviewing applicants' resumes and cover letters, and it is not news that resumes and cover letters are easily faked. Recruiters do not research these candidates properly; they recruit based on what they see without checking for actual experience.

4. Transaction-Oriented: A recruiter is more transaction-oriented. He is only interested in the financial benefits he would gain after recruiting people to a company, and he isn't interested in how good they are or what values they could offer to the company. Let's say this treats recruitment as a "first come, first serve" offer instead of a thoughtful process aimed at getting the perfect Candidate.

5. Limited business knowledge: Recruiters do not understand the company's needs. They do not know the company's challenges or the kind of candidates the company is looking for. The fact that they do not understand this context would lead to them recruiting just anybody.

TALENT ACQUISITION SPECIALIST.

1. Strategically partners with hiring teams: A talent acquisition specialist forms a kind of relationship with the hiring teams or the company to understand their specific needs, goals, and even their challenges. This would help them develop some strategies for finding the perfect Candidate for the job, give them insights to make good decisions, and also ensure a smooth hiring process that meets the company's needs.

2. Often works proactively. Building long-term talent pipelines: Talent acquisition specialists frequently work proactively. They anticipate future talent needs and build on long-term talent pipelines. This helps them to identify potential candidates even before the job positions are available; it helps them develop relationships with top talents to maintain their network of qualified professionals. It helps them stay ahead of the competitors as they always attract or get the best talents.

It ensures a consistent flow of qualified applicants to meet future business needs. It can also save time.

3. Uses data, branding, and Candidate experience as tools: Talent acquisition specialists use data, employer branding, and candidate experience to hire. They do not just read applicants' resumes and cover letters. Instead of just recruiting anybody, they research to find out if the candidates are experienced enough for the job. This also helps showcase the company's unique culture, values, and mission to attract like-minded candidates.

4. Relationship Oriented: Talent acquisition specialists focus on building and maintaining relationships with the company to understand their interests and provide guidance to the candidates by offering personalized experiences, giving feedback and maintaining a level of trust between them. With the company stakeholders by collaborating with HR marketing and other teams to drive business objectives. They put these relationships before monetary gains as they know the importance of maintaining such relationships, and the benefits they could get from it are more than just quick cash.

5. Business Knowledge: Talent acquisition specialists possess a deep understanding of the organization's business goals and objectives, market trends and competitors, products and services, cultures and values. This helps them understand the kind of candidates the company is looking for.

THE RECRUITER TO TALENT PARTNER EVOLUTION LADDER

A visual ladder showing progression.

EVOLUTION LADDER
THE RECRUITER TO TALENT PARTNER

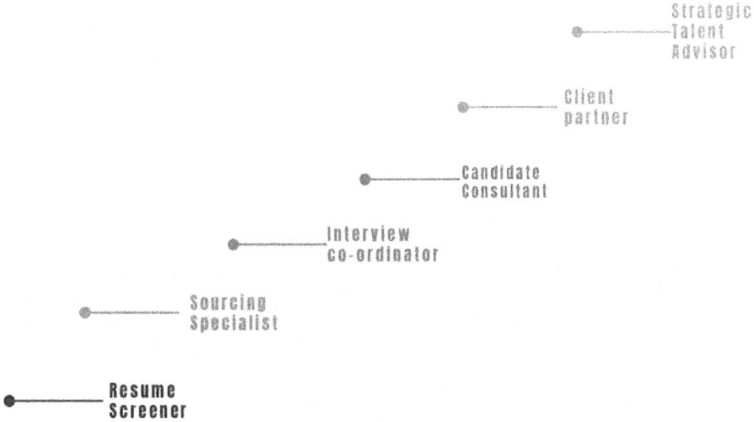

Strategic
Talent
Advisor

Client
partner

Candidate
Consultant

Interview
co-ordinator

Sourcing
Specialist

Resume
Screener

THE TALENT MANAGEMENT CYCLE.

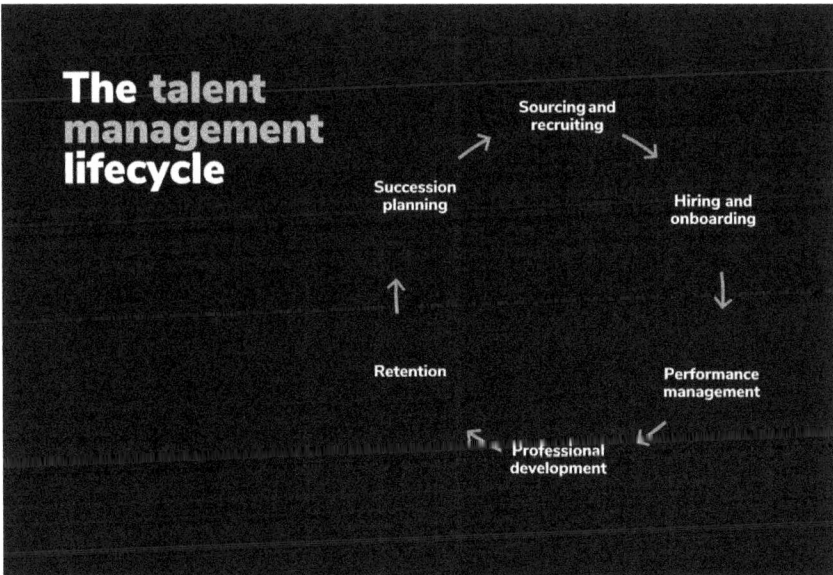

The talent management lifecycle

Sourcing and recruiting

Succession planning

Hiring and onboarding

Retention

Performance management

Professional development

27

1. Resume Screener: The Foundation of Recruitment

The Resume Screener is the entry point of the recruitment process. This role involves reviewing resumes and cover letters to identify potential candidates who will match the job requirements. The Resume Screener's primary focus is filtering out unqualified applicants and presenting suitable candidates to the hiring manager for proper scrutinizing and further selection.

Responsibilities of the Resume screener:

- Reviewing resumes and cover letters
- Conducting initial screenings to assess candidate qualifications
- Identifying potential candidates for further evaluation
- Maintaining accurate records of candidate applications

2. Sourcing Specialist: The Hunter of Hidden Talent

The Sourcing Specialist is also called The Hunter of Hidden Talent because it takes the recruitment process to the next level by proactively seeking out top talent. This role involves utilizing various sourcing strategies, such as social media, job boards, and professional networks, to identify and engage potential candidates. The Sourcing Specialist aims to build a pipeline of qualified candidates for current and future job openings.

Responsibilities of the Sourcing Specialist:

- Developing and implementing sourcing strategies
- Utilizing social media, job boards, and professional networks to identify potential candidates
- Building and maintaining a pipeline of qualified candidates
- Collaborating with hiring managers to understand their talent needs

3. Interview Coordinator: The Master of Logistics

The Interview Coordinator plays a vital role in ensuring a seamless interview process. This role involves coordinating logistics, such as scheduling interviews, arranging travel accommodations, and preparing interview materials. The Interview Coordinator focuses on providing an exceptional candidate experience while supporting the hiring manager's evaluation process.

Responsibilities of the Interview Coordinator:

- Coordinating interview logistics, including scheduling and travel arrangements
- Preparing interview materials, such as resumes and assessment tools
- Ensuring a smooth candidate experience throughout the interview process
- Providing administrative support to hiring managers

4. Candidate Consultant: The Trusted Advisor

The Candidate Consultant takes on a more advisory role, providing guidance and support to candidates throughout the recruitment process. This role involves building relationships with candidates, understanding their career goals and motivations, and presenting job opportunities that align with their aspirations. The Candidate Consultant's goal is to create a positive candidate experience while also ensuring they are the best fit for the company, as the expected results are for both the Candidate and the company to work in alignment.

Responsibilities of the Candidate Consultant:

- Building relationships with candidates and understanding their career goals
- Presenting job opportunities that align with candidates' aspirations
- Providing guidance and support throughout the recruitment process
- Ensuring a positive candidate experience

5. Client Partner: The Strategic Collaborator

The Client Partner is a strategic role that involves collaborating with hiring managers and other stakeholders to understand their talent needs and develop recruitment strategies. This role requires a deep understanding of the company's business goals, industry trends, and market conditions. The Client Partner focuses on delivering tailored recruitment solutions that meet the company's strategic objectives. This is the next stage after the Candidate consultant, as they work together.

Responsibilities of the Client Partner:

- Collaborating with hiring managers to understand their talent needs
- Developing recruitment strategies that align with business goals
- Providing market insights and intelligence to inform recruitment decisions
- Building and maintaining relationships with key stakeholders

6. Strategic Talent Advisor: The Visionary Leader

The Strategic Talent Advisor is the most senior and strategic role in the recruitment function. This role involves providing visionary leadership and expert talent acquisition and management guidance. The Strategic Talent Advisor focuses on developing and implementing talent strategies that drive business growth, innovation, and competitiveness.

Responsibilities of the Strategic Talent Advisor:

- Developing and implementing talent strategies that align with business objectives
- Providing expert guidance on talent acquisition and management
- Collaborating with senior leaders to shape the company's talent agenda
- Driving innovation and best practices in talent acquisition and management.

WHO IS A GLOBAL RECRUITER?

A global recruiter is a recruitment professional who sources, attracts and hires top talent from around the world to fill job openings in various industries and locations. They work with multinational companies, global staffing agencies, or recruitment process outsourcing (RPO) providers to identify and acquire the best talent, regardless of geographical boundaries.

SKILLS OF A GLOBAL RECRUITER

1. Empathy and Communication

Effective communication and empathy are essential skills for global recruiters. They must understand the needs, concerns, and expectations of candidates, hiring managers, and other stakeholders from diverse cultural backgrounds.

- Active listening and clear articulation of thoughts and ideas
- Ability to understand that there are language barriers and communicate complex information simply
- Empathy and understanding of different cultural norms and values. Some cultures do not tolerate some acts, and a global recruiter is supposed to respect that.

2. Adaptability to Different Time Zones & Cultures

Global recruiters must be adaptable and flexible when working with candidates and clients across different time zones and cultures, especially in Asian and African countries where the time differences are many hours apart.

- Ability to adjust to varying work schedules and deadlines
- Understanding of cultural differences and the differences in communication styles
- Willingness to learn about and appreciate diverse cultural practices and traditions.

3. Domain Awareness (IT roles, tools, and trends)

Global recruiters specializing in IT recruitment must understand IT roles, tools, and trends.

- Knowledge of various IT job functions, such as software development, data science, and cybersecurity
- Familiarity with industry-specific tools like ATS, Social media etc., technologies, and platforms
- Understanding of emerging trends and innovations in the IT industry

4. Ethics and Compliance Knowledge

Global recruiters must know and adhere to various laws, regulations, and ethical standards governing recruitment practices.

- Knowledge of labour laws, data protection regulations, and equal employment opportunity laws
- Understanding of company policies and procedures related to recruitment and hiring

- Commitment to upholding ethical standards and promoting diversity, equity, and inclusion

5. Organizational and CRM/ATS Proficiency

Global recruiters must be organized, efficient, and proficient in using various recruitment tools and technologies.

- Ability to prioritize tasks, manage multiple projects, and meet deadlines
- Proficiency in using CRM (Candidate Relationship Management) and ATS (Applicant Tracking System) software
- Understanding of data analytics and reporting tools to track recruitment metrics and KPIs

6. Curiosity, Research Ability, and Salesmanship

Global recruiters must be curious, resourceful, and skilled in sales and marketing techniques to attract top talent and build strong relationships with clients and candidates.

- Curiosity and desire to learn about new industries, technologies, and trends
- Research skills to identify top talent, market trends, and competitor activity
- Sales and marketing skills to promote job opportunities, build brand awareness, and negotiate offers

SALES IN RECRUITMENT

It is necessary and natural that if you do not believe in your Candidate, your client won't either. Sales in recruitment are also a part of the hiring process, and this is something most people do not know. It is not about manipulating the clients or the candidates but rather about showcasing the Candidate's value as a solution to a challenge the client is facing or what the company needs. It is essential that you have confidence in the Candidate; it also means you should have a level of knowledge of the Candidate and the client, which means knowing the client's needs.

Case example: Sales-driven placement.

A recruiter in Europe was struggling to place a DevOps Engineer with a Silicon Valley firm. Two submissions had failed. She shifted her strategy. She didn't do this by changing the candidates and the narrative. She reframed the Candidate's value around a specific business pain: AWS migration. Instead of showcasing a resume, she pitched the Candidate as a solution. The result? The client was impressed and hired within a week.

LESSON: Sales in recruitment are not just about persuasion; they are about solving real problems with the right people. In this case, the recruiter initially struggled to place the DevOps Engineer candidate, but she soon realized what she was doing wrong. She shifted her approach to focus on the Candidate's specific skills and experience in migrating to Amazon Web Services (AWS); she was able to clearly articulate the Candidate's value, address the client's pain points, and build trust and credibility.

As a global recruiter, you are not just selling a job but a career opportunity, a chance for candidates to thrive and grow. When you believe in a product or brand, you are more confident, passionate, and sometimes more persuasive. You can highlight the importance and benefits of being a part of that brand and the challenges they should expect. Geographical boundaries no longer bind the recruitment industry. The rise of remote work, digital communication tools, and social media has made it possible for recruiters to connect with clients and candidates worldwide.

Asia, Africa, Europe, Oceania, and the Americas are not just talent hubs but also home to thousands of skilled recruiters helping US companies fill critical IT roles. This global talent pool has created new opportunities for recruiters to work with clients from diverse industries and backgrounds.

NOTE: This example is only scenario-based for training and learning purposes, helping us understand the scope better. The real scope can be much deeper and more precise.

Think like a talent Advisor, not just a recruiter.

When you believe in your Candidate,

- You speak with confidence and clarity
- You align their strengths to business challenges

- You anticipate objections and already have answers

Clients who feel that conviction tend to trust your judgement even better.

Self-Assessment:

1. What is the main difference between a recruiter and a talent acquisition specialist?
2. Name two key skills that make a recruiter globally effective.
3. Why is empathy important in global IT recruitment?
4. Describe one way in which recruitment involves sales.
5. How can belief in your Candidate impact your success rate?

Reflective Prompt: Think of a role or industry for which you would feel confident recruiting. Why do you believe in it, and how would that confidence show during your screening or client pitch?- From Resume Pushers to Talent Strategists.

CHAPTER TWO

UNDERSTANDING DOMAINS AND SUBDOMAINS IN IT

The objectives of this learning:

1. Understand the importance of domain-specific knowledge in the IT industry
2. Learn how to differentiate between domains and subdomains
3. Gain awareness of the skills and tools typically associated with each domain
4. Learn how to map job roles within the context of client needs
5. Build confidence in recognizing the right Candidate fit for the right domain

THE IMPORTANCE OF DOMAIN-SPECIFIC KNOWLEDGE IN THE IT INDUSTRY

In the rapidly evolving IT industry, domain-specific knowledge has become crucial for the success of professionals and organizations alike.

But then, what exactly is Domain Specific Knowledge?

Domain-specific knowledge refers to the specialized understanding, knowledge, and expertise required to operate effectively within a specific industry or domain, such as healthcare, finance, or manufacturing. This knowledge is essential for IT professionals, enabling them to develop and implement solutions that meet a particular industry's unique needs and challenges.

Why Domain-Specific Knowledge Matters:

1. Understanding Industry-Specific Challenges: Domain-specific knowledge allows IT professionals to understand a particular industry's unique challenges and pain points. This understanding is critical in developing solutions that address these challenges effectively.
2. Developing Relevant Solutions: With domain-specific knowledge, IT professionals can develop solutions tailored to an industry's specific

needs. This ensures that the solutions are relevant, practical, and meet the industry's requirements.

3. Improving Communication with Stakeholders: Domain-specific knowledge enables IT professionals to communicate effectively with stakeholders, including business leaders, customers, and the final users. This improves collaboration, reduces misunderstandings, and ensures solutions meet the industry's needs.

4. Enhancing Credibility and Trust: IT professionals with domain-specific knowledge are likelier to be perceived as credible and trustworthy by industry stakeholders. They demonstrate a deep understanding of the industry's challenges and needs.

5. Staying Competitive: In today's competitive IT landscape, domain-specific knowledge is a key differentiator. IT professionals and organizations with this knowledge are better positioned to attract and retain clients, talent, and market share.

Benefits of Domain-Specific Knowledge:

There are a few benefits to a concept that involves specificity. Some of these include:

Benefits.	Impact.
Increased Efficiency	Faster screenings, fewer mismatches.
Improved Quality	Solutions and candidates aligned to actual industry needs.
Enhanced Innovation	Insightful, domain-aware problem-solving.
Better Risk Mitigation	Anticipate regulatory or compliance risks (e.g. HIPAA, PCI-DSS)
Client Satisfaction	Deliver results that meet expectations and exceed relevance.

1. Increased Efficiency: Domain-specific knowledge enables IT professionals to develop solutions more efficiently. They are familiar with the industry's specific requirements and challenges, which allows them to get straight to the point.

2. Improved Quality: Solutions developed with domain-specific knowledge are more likely to meet the industry's quality standards and requirements

as the IT professional is going about just one thing, so he would ensure it has the best quality.

3. Enhanced Innovation: Domain-specific knowledge can lead to innovative solutions that address specific industry challenges and needs.
4. Better Risk Management: IT professionals with domain-specific knowledge are better equipped to identify and manage risks associated with industry-specific challenges and regulations. For example, an IT professional dealing with finance is more likely to understand its risks than an IT professional in healthcare.
5. Improved Customer Satisfaction: Solutions developed with domain-specific knowledge are more likely to meet customer needs and expectations, improving customer satisfaction.

TABLE ILLUSTRATING THE CONNECTION BETWEEN RECRUITER DOMAIN KNOWLEDGE—BETTER INTAKE–TARGETED SOURCING–STRONGER HIRES–BUSINESS VALUE

Recruiter domain knowledge	Better intake	Targeted sourcing	Stronger hires	Business value
Understanding of industry trends	Accurate job requirements	Effective job posting	Qualified candidates	Reduced time-to-hire
Knowledge of job roles and responsibilities	Clear candidates profiles	Targeted candidate sourcing	Better cultural fit	Improved employee retention.
Familiarity with required skills and qualifications	Precise candidate evaluation	Efficient screening process	Higher quality hires	Increased productivity
Awareness of market conditions.	Informed recruitment strategies	Competitive job offers	Top talent attraction	Enhanced competitiveness
Insights into Candidate Motlvations	Personalized candidate engagement	Stronger candidate relationships	Improved candidate experience	Positive employer brand

WHY SOURCING IN RECRUITMENT IS IMPORTANT.

Sourcing In Recruitment

Why is it important?

01	02	03	04
Efficiently Attracting High-Quality Candidates	Improvement Based on Data-Driven Insights	Aligning with Organization's Goals	Enhancing Employer Brand

HOW DOES SOURCING WORK IN RECRUITMENT?

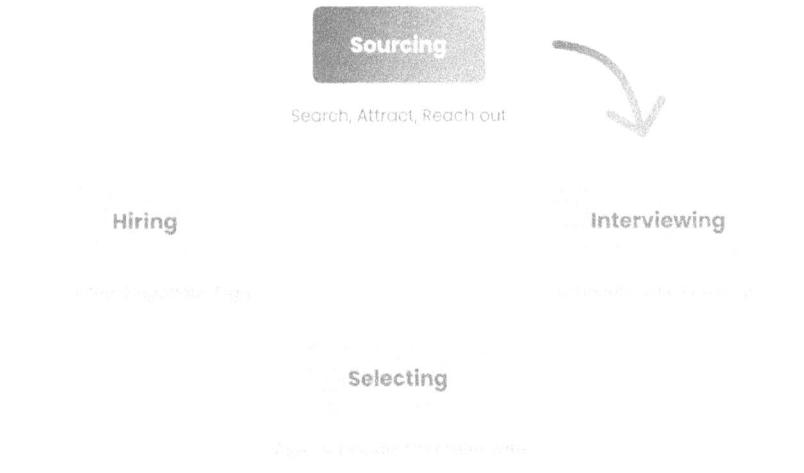

Sourcing

Search, Attract, Reach out

Hiring

Interviewing

Selecting

Examples of domain-specific Knowledge

We had listed some of these examples earlier, but now, we will talk more about them. They include:

1. Healthcare: Understanding healthcare regulations, such as the Health Insurance Portability and Accountability Act (HIPAA), and industry-specific challenges, such as patient data management. It involves everything that has to do with the healthcare sector.
2. Finance: Knowledge of financial regulations, such as the Payment Card Industry Data Security Standard (PCI-DSS), and industry-specific challenges, such as risk management and compliance.
3. Manufacturing: Understanding manufacturing processes, industry-specific challenges, such as supply chain management, and regulations, such as ISO 9001. ISO 9001 is an international standard for quality management systems. It provides a framework for organizations to demonstrate their commitment to quality, customer satisfaction, and continuous improvement.
4. Retail: Knowledge of retail industry trends, such as e-commerce and omnichannel retailing, and industry-specific challenges, such as inventory management and customer experience.

We have repeatedly mentioned the word "industry specific" in this chapter. So, let's explain what it means. Industry-specific means something tailored or relevant to a particular sector or industry. It might include languages, laws and regulations, skills, tools, methodologies, etc.

So we know what that is now. Let's continue talking about Domain Specific Knowledge.

Ways of Acquiring Domain-Specific Knowledge:

For something specific, there are also particular ways it can be acquired. Some of these are mentioned below:

1. Training and Certification: IT professionals can acquire domain-specific knowledge through training and certification programs offered by industry associations, vendors, and educational institutions.
2. Industry Events and Conferences: Attending industry events and conferences can provide IT professionals with opportunities to learn from industry experts and network with peers.

3. Industry Publications and Research: Reading industry publications and research papers can help IT professionals stay current with industry trends and challenges.
4. Mentorship and Coaching: Working with experienced professionals with domain-specific knowledge can provide valuable insights and guidance.
5. On-the-job Experience: IT professionals can acquire domain-specific knowledge through on-the-job experience, such as working on projects and solving industry-specific challenges.

PRO TIP: Add industry-specific RSS feeds to your inbox. Start each day with a ten-minute skim of headlines in healthcare, fintech, cloud, or your target market.

DIFFERENCES BETWEEN DOMAIN AND SUBDOMAIN

Domains and subdomains refer to different levels of specialization in industries and fields of expertise. Although they are somewhat similar, these two concepts are distinct. People who know little about both might not understand this so that we will break it down.

DOMAIN:

A domain is a broad field or industry encompassing various topics, skills, and expertise. Examples of domains include:

1. Healthcare: This refers to the maintenance or improvement of physical and mental health through the prevention, diagnosis, treatment and management of illnesses, diseases and injuries. It is a broad field that includes various specialities, such as medicine, nursing, and allied health professions.
2. Technology: Technology refers to applying scientific knowledge, tools, and techniques to create, develop, and use machines, devices, systems, and methods to solve problems, improve efficiency, and enhance human life. It is a domain that encompasses various subfields, such as software development, data science, and cybersecurity.
3. Finance: Finance refers to the management of money, investments, and financial resources, including activities such as banking, investing, lending, and financial planning. The goal is to achieve economic stability,

growth, and profitability. It is a domain that includes various specialities, such as accounting, investment banking, and asset management.

4. Education: This refers to the process of teaching, learning, and acquiring knowledge, skills, and values through various methods, including formal schooling, training, and personal development, with the goal of personal growth, intellectual development, and social mobility. It is a domain that includes various specialities, such as teaching, curriculum development and educational administration.

5. Environmental Science: This is the study of the natural world and the impact of human activities on the environment, including the air, water, soil, and living organisms, with the goal of understanding and addressing environmental issues, such as pollution, climate change, and conservation. A domain that encompasses various subfields, such as ecology, conservation biology, and environmental policy.

6. Engineering: Engineering is the application of scientific and mathematical principles to design, develop, and test structures, machines, systems, and processes to solve practical problems, improve efficiency, and enhance human life. It is a domain that includes various specialities, such as mechanical engineering, electrical engineering, and civil engineering.

7. Arts and Design: Arts and design refer to the creative expression and application of imagination, skill, and technique to produce works of art, music, literature, and visual and performing arts, as well as the design of products, spaces, and experiences that combine aesthetics, functionality, and creativity.A domain that encompasses various subfields, such as graphic design, fine arts, and music.

8. Social Sciences: Social science studies human behaviour, relationships, and institutions, including psychology, sociology, anthropology, economics, and politics. Its goal is to understand and explain social phenomena, behaviours, and interactions.

9. Business and Management: Business management refers to planning, organizing, leading, and controlling an organization's resources, including people, finances, and operations, to achieve its goals, objectives, and mission and maximize efficiency, productivity, and profitability. It encompasses various subfields, such as marketing, finance, and Human Resources.

10. Computer Science: Computer science studies computers, programming languages, and algorithms, including the design, development, and testing of computer systems, software, and applications to solve problems, process information, and improve human-computer interactions. This domain includes various specialities, such as software development, data science, and artificial intelligence.
11. Health and Wellness: Health and wellness refer to physical, mental, and emotional well-being, encompassing nutrition, fitness, stress management, and self-care. The goal is optimal health, happiness, and quality of life. This domain covers various subfields, such as nutrition, fitness, and public health.
12. Law and Justice: Law and justice refer to the system of rules, regulations, and institutions that govern society, protect individual rights, and ensure fairness, equality, and accountability. They aim to maintain social order, resolve disputes, and uphold the rule of law. This domain includes various specialities, such as criminal, civil, and international law.
13. Agriculture: Agriculture refers to the cultivation of crops, the raising of livestock, and the production of food, fibre, and other products using various techniques, technologies, and practices. Its goal is to provide sustenance, support rural development, and promote food security. It encompasses multiple subfields, such as agronomy, horticulture, and animal science.

SUBDOMAIN:

A subdomain is a more specific area of expertise within a broader domain. Subdomains are often more specialized and require more profound knowledge and expertise.

Examples of subdomains include:

1. Cardiology (a subdomain of healthcare): A specialized field that focuses on diagnosing, treating, and preventing heart and blood vessel disorders.
2. Artificial intelligence(a subdomain of technology): A specialized field that focuses on developing intelligent systems that can perform tasks that typically require human intelligence.

3. Financial analysis (a subdomain of finance): A specialized field that focuses on studying financial data to inform business decisions.
4. Special Education (a subdomain of education): A specialized field that focuses on teaching students with special needs.
5. Renewable Energy (a subdomain of environmental science): A specialized field focusing on developing sustainable energy sources.
6. Aerospace Engineering (a subdomain of engineering): A specialized field that focuses on designing and developing aircraft and spacecraft.
7. Graphic Design (a subdomain of arts and design): A specialized field that focuses on creating visual content for various media.
8. Clinical Psychology (a subdomain of social sciences): A specialized field that focuses on assessing and treating mental health disorders.

DIAGRAM: A VISUAL TREE SHOWING DOMAINS AND SUBDOMAINS

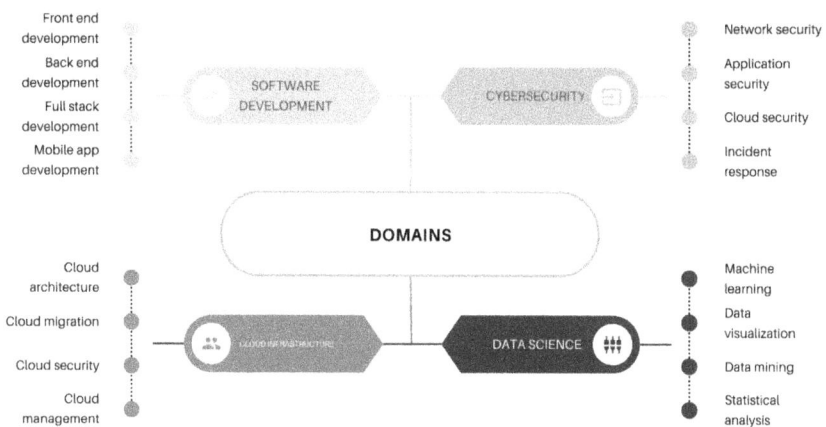

DIFFERENCES BETWEEN DOMAIN AND SUBDOMAIN

Domain vs Subdomain in Industries and Fields of Expertise

Attribute	Domain	Sub Domain
Scope	Broad	Narrow and specific
Knowledge level	General understanding	Specialized expertise
Example	Cybersecurity	Application security, GRC, Pen testing
Use in intake	Identifies overall area of need	Determines precise skills/tools required
Use in sourcing	Filters target Candidate	Optimizes keyword/skill search
Example Job titles	Security Engineer	SOC Analyst, IAM Specialist, Compliance Lead

We are going to be elaborating further on these differences:

DOMAIN:

1. Broad field of expertise: A domain refers to a broad field or industry, such as healthcare or finance.
2. General knowledge area: Domains represent general areas of knowledge or practice.
3. High-level categorization: Domains are high-level categories that encompass various subfields. A domain can be described as a tree, with the subdomains being its branches.
4. Overarching theme: Domains often have an overarching theme or focus. An overriding theme or focus is the central idea or objective that ties together various aspects, elements, or components of a particular domain. For example, the overarching theme in healthcare could be "to improve patient outcomes." This would guide the development of subdomains under healthcare.
5. Wide scope: Domains typically have a broad scope, covering many aspects of a field.

SUBDOMAIN:

1. Specialized field: A subdomain is a specialized field within a broader domain.
2. Narrower focus: Subdomains have a narrower focus, such as cardiology within healthcare.
3. Specific application: Subdomains often involve specific applications or techniques.
4. Subset of knowledge: Subdomains represent subsets of knowledge or expertise within a domain.
5. More detailed: Subdomains are more detailed and specific than domains.

Additional Differences between Domain and Subdomain:

1. Level of specialization: Domains are more general, while subdomains are more specialized.
2. Scope of practice: Domains have a broader scope of practice, while subdomains have a narrower scope.
3. Expertise required: Domains may require general knowledge, while subdomains require specialized expertise.
4. Industry recognition: Domains are often widely recognized, while subdomains may be more niche.
5. Training and education: Domains may require general education, while subdomains often require specialized training or certifications.

WHY DOES THIS DISTINCTION MATTER?

Understanding the distinction between domain and subdomain is crucial because it helps you:

- Source accurately: Identify the right skill level and focus for specific job requirements.
- Communicate effectively: Confidently speak the client's language, building trust and credibility.
- Optimize sourcing: Create targeted Boolean strings to find the best candidates.
- Evaluate resumes precisely: Assess resumes in context, beyond just keywords, to ensure a better fit.

RECRUITER SCENARIO: APPLIED INSIGHT

Let's say a client asks you for a DATA SCIENTIST, but after probing, you realize they need someone with deep experience in NATURAL LANGUAGE PROCESSING (NLP) to work on chatbot analytics. That is a subdomain of AI/ML, not general data science. Without knowing this distinction, you might submit results that are not even close to the mark. With subdomain fluency, you instantly pivot and source profiles with tools like Python + spaCy or NLTK– and wow the client with precision.

SKILLS AND TOOLS ASSOCIATED WITH SOME OF THE DOMAINS LISTED ABOVE

Arts and Design

- Skills: Creativity, visual thinking, attention to detail, communication.
- Tools: Adobe Creative Cloud (Photoshop, Illustrator, InDesign), Sketch, Figma, graphic design software.

Business Management

- Skills: Leadership, strategic planning, problem-solving, financial analysis.
- Tools: Microsoft Office, Google Workspace, project management software (Asana, Trello), accounting software (QuickBooks).

Computer Science

- Skills: Programming, data structures, algorithms, software engineering.
- Tools: Programming languages (Java, Python, C++), development environments (Eclipse, Visual Studio), version control systems (Git).

Education

- Skills: Teaching, curriculum design, assessment, communication.
- Tools: Learning management systems (Canvas, Blackboard), educational software (Kahoot), online resources (Coursera, edX).

Engineering

- Skills: Problem-solving, mathematical modelling, design, prototyping.
- Tools: Computer-aided design (CAD) software (Autodesk, SolidWorks), simulation tools (ANSYS), programming languages (MATLAB, Python).

Environmental Science

- Skills: Scientific research, data analysis, environmental policy, sustainability.
- Tools: Geographic information systems (GIS), remote sensing software, environmental monitoring equipment.

Health and Wellness

- Skills: Patient care, medical knowledge, communication, empathy.
- Tools: Electronic health records (EHRs), medical imaging software, fitness tracking devices.

Law and Justice

- Skills: Legal research, critical thinking, communication, advocacy.
- Tools: Legal databases (Westlaw, LexisNexis), case management software, courtroom technology.

Social Science

- Skills: Research, data analysis, critical thinking, communication.
- Tools: Statistical software (SPSS, R), qualitative data analysis software (NVivo), online survey tools (SurveyMonkey).

Agriculture

- Skills: Crop management, animal husbandry, agricultural engineering, sustainability.
- Tools: Precision agriculture software, farm management software, agricultural equipment.

GRID HIGHLIGHTING CRITICAL SKILLS FOR SUBDOMAINS

IT Domain	Sub domain	Critical tools/skills
Software development	Front-end, javascript, React, HTML/CSS, Back-end, java, Python, node	Combination of front-end and back-end skills
Cybersecurity	Network security, firewall configuration, VPN.	Application security, secure coding practices, vulnerability assessment, cloud security, cloud security architecture, compliance.
Cloud infrastructure	Cloudarchitecture,AWS, Azure, Google cloud	Cloud migration, cost optimization, cloud management, cloud monitoring, cost management
Data Science	Machine learning, Python, R TensorFlow	Data visualization, Tableau, power BI,D3.js, statistical analysis, data modelling.

MAPPING JOB ROLES TO MATCH CLIENTS NEEDS

When mapping out job roles, we must consider the client's needs. Mapping out roles when a client or clients do not need it is sometimes considered a waste. Therefore, To map job roles to client needs, follow these steps:

1. Identify Client Needs:

- Determine the client's goals, objectives, and challenges.
- Understand their industry, market, and target audience.
- Identify the specific pain points or problems they need to solve.

2. Define Job Roles:

- Determine the job roles required to address the client's needs.
- Consider the skills, expertise, and responsibilities required for each role.
- Define the key performance indicators (KPIs) for each role.

3. Map Job Roles to Client Needs:

- Create a matrix or table to map job roles to client needs.
- Identify the specific job roles that align with each client's needs.
- Consider the skills, expertise, and responsibilities required for each role.

4. Consider the Client's Organization:

- Understand the client's organizational structure and culture.
- Identify the key stakeholders and decision-makers.
- Consider the client's existing technology infrastructure and systems.

5. Validate the Mapping:

- Validate the mapping with the client to ensure it meets their needs.
- Refine the mapping as necessary to ensure accuracy and relevance.

Example of mapping:

CLIENT NEED	JOB ROLE	SKILLS/EXPERTISE	RESPONSIBILITIES
To improve customer experience.	Customer experience manager.	Customer journey mapping, UX design.	Develop customer experience strategy and implement improvements.
Increase sales.	Sales analyst.	Data analysis and sales forecasting.	Analyze sales data, identify trends, and develop sales strategies.
Enhance Cyber security.	Cyber security engineer.	Network security, threat analysis.	Design and implement security measures and monitor for threats.
Implement Cloud solutions.	Cloud engineer.	Cloud architecture, migration planning.	Design and implement Cloud solutions and migrate applications.

By following these steps and creating a mapping of job roles to client needs, you can ensure that the right talent is aligned with the client's goals and objectives.

BUILDING CONFIDENCE IN RECOGNIZING THE RIGHT CANDIDATE FIT

As each domain has its specificity, it is the same way there are different candidates for each domain. But then, it is another thing to build the confidence in recognizing the right Candidate. To build confidence in recognizing the right Candidate fit for the right domain, follow these steps:

1. Develop Domain Expertise:

- Study the domain and its specific requirements.
- Stay up to date with industry trends and developments.
- Network with professionals in the domain to gain insights.

2. Define Clear Job Requirements:

- Create detailed job descriptions outlining key responsibilities and skills.
- Identify essential qualifications, experience, and education.
- Consider soft skills and cultural fit.

3. Use Relevant Assessment Tools:

- Utilize skills assessments, personality tests, and cognitive evaluations.
- Leverage tools like behavioural interviews, case studies, or presentations.
- Consider using AI-powered assessment platforms.

4. Evaluate Candidate Experience:

- Review resumes, cover letters, and online profiles.
- Conduct thorough interviews, asking behavioural and situational questions.
- Assess candidate projects, portfolios, or writing samples.

5. Validate Assumptions:

- Verify candidate claims through reference checks or background screening.
- Validate skills and knowledge through practical assessments.
- Consider using trial or pilot projects.

6. Continuously Refine Your Process:

- Reflect on past hiring decisions and outcomes.
- Gather feedback from candidates, hiring managers, and peers.
- Adjust your approach as needed to improve accuracy.

Key Indicators of a Good Fit

When evaluating candidates, look for the following:

1. Relevant experience: Direct experience in the domain or related fields.
2. Transferable skills: Skills that can be applied to the domain, even if not directly related.
3. Domain knowledge: Understanding the domain's concepts, trends, and challenges.
4. Problem-solving skills: Ability to analyze problems and develop practical solutions.
5. Cultural fit: Alignment with the organization's values, mission, and work environment.

By following these steps and considering key indicators, you'll build confidence in recognizing the right Candidate fit for the right domain.

PRO TIP: A candidate who doesn't know the tool but understands the problem it solves is often more adaptable than someone who only knows how to 'click buttons.'

WHY DOMAIN KNOWLEDGE MATTERS WHETHER YOU COME FROM A TECHNICAL BACKGROUND OR NOT

Earlier in this chapter, we discussed why domain-specific domain-specific domain-specific knowledge matters; now, we will discuss why domain issues of expertise, regardless of one's technical background.

Domain knowledge is essential because it provides the following:

1. Context: Understanding the industry, its challenges, and its specific needs.
2. Relevance: Knowing how to apply skills and knowledge to real-world problems.
3. Communication: Ability to effectively communicate with stakeholders, clients, or customers.

4. Problem-solving: Familiarity with domain-specific challenges and solutions.
5. Innovation: Understanding the nuances of the domain nuances of the domain to drive innovation and improvement.

The Value of Domain Knowledge in Recruitment:

In the world of recruitment, it's often said that hiring managers do not just want resumes. Instead, they want solutions to particular problems. This couldn't be more true, especially in the tech industry, where the shades between different roles and specializations can be vast. As a recruiter, understanding these differences is crucial to becoming a strategic partner rather than just a resume collector.

The Importance of Domain Knowledge:

Domain knowledge is the foundation upon which successful recruitment is built. When you have a deep understanding of the specific needs and challenges of a particular industry or role, you're able to identify top talent more effectively. This isn't just about knowing the buzzwords or the latest trends; it's about having a genuine comprehension of what makes a candidate a good fit for a specific position.

Cloud Engineer vs. Cybersecurity Analyst: Understanding the Difference

For instance, consider the difference between a cloud engineer and a cybersecurity analyst. Both roles are critical in today's tech landscape, but they require vastly different skill sets and mindsets.

- A cloud engineer is responsible for designing, building, and maintaining cloud computing systems. They need to have expertise in areas such as cloud architecture, migration, and management.
- A cybersecurity analyst, on the other hand, is focused on protecting an organization's digital assets from cyber threats. They need to have a deep understanding of security protocols, threat analysis, and incident response.

Without domain knowledge, it would be challenging to distinguish between these two roles and identify the right candidate for each position.

SaaS Implementation vs. Custom Software Development: Understanding the Different Shades.

Similarly, understanding the different shades between SaaS implementation and custom software development is essential for identifying the right talent.

- SaaS implementation involves deploying and configuring software applications over the internet. It requires expertise in areas such as software integration, data migration, and user adoption.
- Custom software development, on the other hand, involves creating bespoke software solutions tailored to an organization's specific needs. It requires expertise in areas such as software design, development, and testing.

By understanding these differences, recruiters can identify candidates who have the specific skills and experience required for each role.

Domain Overviews and Examples (In tech).

1. Software Development:

- Subdomains: Frontend (React, Angular), Backend (Java, .NET, Python), Full Stack, Mobile (iOS, Android)
- Tools: Git, GitHub, Jenkins, IntelliJ, Visual Studio
- Sample Job Titles: Java Developer, Full Stack Engineer

2. Quality Assurance (QA):

- Subdomains: Manual Testing, Automation Testing, Performance Testing
- Tools: Selenium, JUnit, LoadRunner, Postman
- Sample Job Titles: QA Analyst, Test Automation Engineer

3. Cybersecurity:

- Subdomains: Network Security, Application Security, Governance, Risk & Compliance (GRC), Penetration Testing.
- Tools: Splunk, Wireshark, Nessus, Palo Alto, CrowdStrike.
- Sample Job Titles: SOC Analyst, Security Engineer.

4. Cloud & DevOps:

- Subdomains: AWS, Azure, GCP, Infrastructure as Code (IaC), CI/CD Pipelines.
- Tools: Terraform, Kubernetes, Docker, Jenkins.
- Sample Job Titles: Cloud Engineer, DevOps Specialist.

5. Data Science & AI:

- Subdomains: Machine Learning, NLP, Data Engineering, Visualization
- Tools: Python, R, Tableau, Power BI, TensorFlow, PyTorch
- Sample Job Titles: Data Scientist, ML Engineer

6. SaaS & Enterprise Software:

- Subdomains: Implementation, Integration, Product Management
- Tools: Salesforce, SAP, Oracle, NetSuite
- Sample Job Titles: SaaS Consultant, Product Manager

7. Healthcare IT:

- Subdomains: Claims, EMR/EHR Systems, Compliance, Patient Portals
- Tools: Epic, Cerner, HealthEdge, Facets
- Sample Job Titles: Business Systems Analyst (Healthcare), Claims Config Analyst

8. Telecom & FinTech:

- Subdomains: Billing Systems, Payment Gateways, Compliance, Core Banking Systems
- Tools: Amdocs, FIS, Finastra, Swift, ISO20022
- Sample Job Titles: Telecom Engineer, FinTech Solutions Architect

These domains and sub-domains represent various areas of expertise in the tech industry, with corresponding tools and job titles. Each domain requires specific skills and knowledge, and professionals can specialize in one or multiple areas.

Becoming a Strategic Partner:

When you have domain knowledge, you're able to position yourself as a strategic partner rather than just a recruiter. You're able to understand the specific needs and challenges of the hiring manager and identify top talent that meets those needs.

This involves:

- Developing a deep understanding of the industry and the specific roles and specializations within it.
- Building relationships with hiring managers and understanding their specific needs and challenges.
- Identifying top talent and presenting them with candidates who have the specific skills and experience required for each role.

By becoming a strategic partner, recruiters can add more value to the hiring process and help organizations find the best talent to drive their business forward.

Domain/Subdomain matrix:

DOMAIN/SUB DOMAIN MATRIX

Domain	Sub Domain	Sample Job title	Tools/ Technol ogies
Software Development	Backend(Java)	Java developer	IntelliJ, Github
Quality Assurance	Automation	Test automation engineer	Selenium, J Unit
Cybersecurity	SOC operations	SOC Analyst	Splunk, wireshark
Arts&Design	Editing	Photo editor	photoshop
Computer Science	Programming	Computer Programmer	Python

DOMAIN MODEL FOR THE MODELING AND ANALYSIS OF CLINICAL PATHWAYS

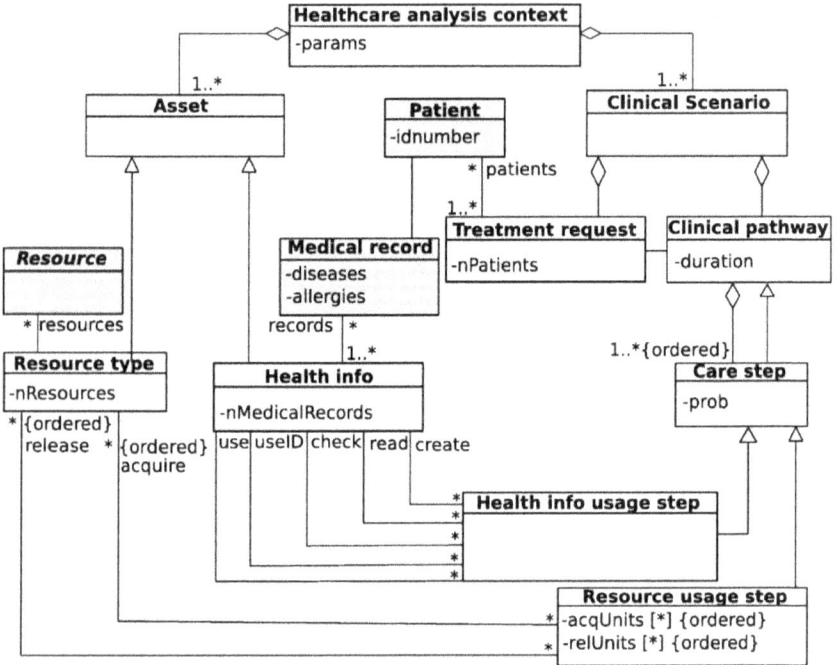

HOW TO USE THIS KNOWLEDGE

1. Asking Domain-Specific Questions During Intake Calls:

To effectively use domain knowledge during intake calls, ask questions that demonstrate your understanding of the client's specific needs and challenges. This will help you:

- Build credibility: Show the client that you are knowledgeable about their industry and role.

Gather information: Collect specific details about the client's requirements and pain points.

- Identify key skills: Determine the essential skills and qualifications required for the role.

Example Questions

- What are the biggest challenges facing your team in the [specific domain], and how do you see this role contributing to solving them?
- Can you walk me through the technical requirements for this position, including any specific tools or technologies?
- How does this role fit into your overall [industry/field] strategy, and what are the key performance indicators (KPIs) for success?
- What sets this role apart from similar positions in your organization, and what unique qualities are you looking for in a candidate?
- Are there any specific pain points or areas of improvement that you'd like the new hire to address, and how will their success be measured?

2. Matching Resumes to Domain Vocabulary While Sourcing:

To effectively match resumes to domain vocabulary while sourcing, consider the following strategies:

Understanding Domain Vocabulary

- Familiarize yourself with industry terminology: Learn the specific words, phrases, and acronyms used in the domain.
- Identify key skills and technologies: Determine the essential skills, tools, and technologies required for the role.

Resume Screening

- Use domain-specific keywords: Search for resumes containing relevant keywords and phrases.
- Look for relevant experience: Identify candidates with experience in the specific domain or industry.
- Assess technical skills: Evaluate candidates' technical knowledge and proficiency in specific tools and technologies.

Example Of vocabularies in some domains:

- Cloud computing: AWS, Azure, Google Cloud, cloud architecture, migration, management.
- Cybersecurity: Security protocols, threat analysis, incident response, penetration testing, security compliance.

- Data science: Machine learning, data analytics, data visualization, statistical modeling, data mining.

3. Aligning Skills/Tools with Sub Domain Requirements in Screening:

- To effectively align skills/tools with subdomain requirements in screening, consider the following strategies:

Understanding Sub Domain Requirements

- Identify specific subdomain subdomain requirements: Determine the unique skills, tools, and technologies required for the subdomain subdomain.
- Familiarize yourself with subdomain terminology: Learn the specific words, phrases, and acronyms used in the subdomain.

Screening Criteria

- Develop sub-domain-specific screening criteria: Create a checklist of essential skills, tools, and technologies required for the subdomain.
- Assess candidate experience: Evaluate candidates' experience working in the subdomain or with similar technologies.
- Evaluate technical skills: Assess candidates' technical knowledge and proficiency in specific tools and technologies.

Example Sub Domains and Required Skills

- Cloud Computing - AWS: Experience with AWS services such as EC2, S3, Lambda, and CloudFormation.
- Cybersecurity - Penetration Testing: Knowledge of penetration testing tools such as Metasploit, Burp Suite, and Nmap.
- Data Science - Machine Learning: Experience with machine learning algorithms, libraries, and frameworks such as TensorFlow, PyTorch, and sci-kit-learn.

4. Talking in Terms Clients Understand

When discussing candidates with clients, it's essential to communicate in terms they understand. This means using specific language and terminology relevant to their industry or domain.

Case Example: Cloud Engineer

Suppose you're tasked with finding a cloud engineer, and you have two candidates:

- Candidate A has experience with AWS Lambda, Terraform, and Jenkins.
- Candidate B is Azure certified with manual deployment experience.

The job specification requires experience with Infrastructure as Code (IaC).

Recommendation

You recommend Candidate A with confidence because "He has GCP experience with Terraform and Docker for IaC." You are confident of this because you know this candidate's qualifications, and you are sure of it. This builds your credibility by:

- Demonstrates understanding: By using specific terminology such as IaC, Terraform, and Docker, you indicate that you understand the client's needs and requirements.
- Relevant experience: You highlight Candidate A's relevant experience with AWS Lambda, Terraform, and Jenkins, which aligns with the job specification.
- Confidence in recommendation: By recommending Candidate A with confidence, you demonstrate that you've thoroughly evaluated the candidates and are committed to finding the best fit for the client's needs.
- Technical expertise: Your explanation showcases your technical knowledge and ability to assess candidates' skills and experience.

NOTE: This example is scenario-based for training and learning purposes to help us understand the scope better. The real scope can be more in-depth and precise.

Self-Assessment:

1. What is the difference between a domain and a sub-domain in IT?
2. Name two sub-domains within cybersecurity.
3. Which tools are associated with Automation Testing?
4. What domain does Terraform typically belong to?
5. Why is it essential to understand domain-specific tools during screening?

Reflective Prompt: Choose one domain you are least familiar with. Research two job roles and one tool used in that area. How could you start building confidence in discussing those roles with candidates or clients?- Why Domain Knowledge Matters (Even for Non-Tech Recruiters)

CHAPTER THREE

ORGANIZATIONAL STRUCTURES AND ROLE FITMENT

The objectives of this learning:

- Understand how IT teams are structured in U.S. companies
- Identify who's involved in the hiring decision and why their roles matter
- Learn to read and use IT org charts during sourcing and intake discussions
- Recognize where different IT roles sit within technical and business teams
- Build confidence in mapping candidates to team structures and client needs

The Information Technology Team Structure in the U.S.

IT teams in the U.S. can vary in structure depending on the organization, industry, and size. However, here is a general overview of common IT team structures:

There are three primary IT team structures in the U.S.: the Centralized IT team, the Decentralized IT team, and the Hybrid IT team. Each team comprises its own characteristics, which will be talked about below:

Centralized IT Team: A centralized IT team is a single team that handles all IT functions for an organization, with a clear hierarchy and defined roles and responsibilities. This structure enables economies of scale, shared resources, and expertise but may limit autonomy and flexibility for individual departments or locations.

Characteristics of a Centralized IT team. Some of these characteristics were stated in the definition but will be explained in short notes below.

1. Single team: A single team handling all IT functions means that one centralized team is responsible for all IT-related tasks, services, and

support across the organization, providing a unified and cohesive approach to IT management.

2. Clear hierarchy: Defined roles and responsibilities. A clear hierarchy means defined roles and responsibilities, with established reporting lines and a chain of command, ensuring accountability, efficiency, and effective decision-making.

3. Economies of scale: Shared resources and expertise. The resources are shared among all team members, and they are also mostly experts in their fields.

Decentralized IT Team: A decentralized IT team is a structure where separate IT teams are established for different departments, locations, or divisions within an organization, allowing for more autonomy and flexibility but potentially leading to duplication of efforts and resources.

Characteristics of a Decentralized IT team.

1. Multiple teams: Separate IT teams for different departments or locations.

2. Autonomy: Teams have more control over their IT functions. The teams have more control over their IT functions, allowing for greater flexibility, decision-making power, and responsiveness to specific departmental or location needs without interference from others.

3. Potential for duplication: Decentralized IT teams may lead to overlapping efforts and resources, resulting in duplicated work, wasted resources, and inefficiencies.

Hybrid IT Team: A hybrid IT team combines elements of centralized and decentralized structures, balancing flexibility and autonomy with shared resources and expertise.

Characteristics of a hybrid IT team:

Just like in the centralized, the characteristics of Hybrid had also been said in its definition, but they would be further explained below:

1. Combination of centralized and decentralized: Some IT functions are centralized, while others are decentralized.

2. Flexibility: Balances the benefits of both centralized and decentralized structures.

3. Complexity: A Hybrid IT structure can be more challenging to manage due to the combination of centralized and decentralized elements, requiring careful coordination and communication to ensure seamless operations.

PROS, CONS, AND BEST FITS FOR EACH STRUCTURE

Team structure	Pros	Cons	Best fit scenarios
Centralized	Standardization, efficiency, easier management,	Limited flexibility and slow response to departmental needs.	Large, stable organizations with standardized processes
Decentralized	Faster responses to departmental needs, more flexible.	Duplication of efforts, lack of standardization.	Dynamic, departmentalized organizations with unique needs.
Hybrid	Balances standardization and flexibility, leverages expertise	Complexity, potential for conflict	Large, complex organizations with diverse needs and requirements.

WHY UNDERSTANDING A CLIENT'S STRUCTURE HELPS:

Understanding a client's IT Structure helps you:

- Ask better intake questions.
- Identify the correct decision-makers
- Assess team fit and reporting dynamics
- Reduce mismatches and offer dropouts
- Customize your candidate pitch

Common IT Roles:

1. IT Director/Manager: The IT Director/Manager oversees the entire IT function, developing and implementing strategies, managing budgets, and ensuring alignment with organizational goals.
2. Network Administrator: The Network Administrator manages and maintains computer networks, ensuring their security, performance, reliability, and troubleshooting issues as they arise.
3. Systems Administrator: Systems Administrators manage, maintain, and support an organization's computer systems, including servers,

databases, and applications, ensuring their security, performance, and reliability.

4. Cybersecurity Specialist: Focuses on security and risk management. Cybersecurity Specialists protect an organization's computer systems and data from cyber threats by implementing security measures, monitoring for vulnerabilities, and responding to incidents.

5. Help Desk Technician: Provides technical support. Help Desk Technicians provide technical support and assistance to users, troubleshooting and resolving hardware, software, and network issues.

6. Software Developer: Software Developers design, create, and maintain software applications, writing code, testing, and debugging programs to meet user needs and requirements.

7. Data Analyst: Data Analysts collect, analyze, and interpret data to identify trends, create visualizations, and inform business decisions, helping organizations make data-driven choices.

Industry Specific IT Teams:

1. Finance: May have separate teams for trading platforms, risk management, and compliance.

2. Healthcare: May have teams focused on medical records, billing, and patient engagement.

3. E-commerce: We may have teams focused on online platforms, digital marketing, and customer experience.

Trends in IT Team Structure:

1. Cloud computing: Increasing adoption of cloud services.

2. DevOps: Emphasis on collaboration between development and operations teams.

3. Cybersecurity: Growing importance of security and risk management.

4. Artificial intelligence: Integration of AI and machine learning into IT operations.

DIAGRAM: WHAT A TYPICAL IT TEAM LOOKS LIKE:

HIRING DECISIONS IN IT.

People involved in Hiring Decisions and why their roles matter:

There are multiple people or offices involved in hiring new workers for a company. These offices play a crucial role in the company's success by hiring workers who align with the company's goals. Some of them include:

1. Hiring Manager: Responsible for defining job requirements and evaluating candidate fit.
2. Recruiter: Sources and screens candidates to ensure a smooth hiring process.
3. Department Heads: Provide input on team needs and validate the skills of candidates.

4. HR Representative: Ensures compliance with company policies and procedures.
5. Team Members: May participate in interviews to assess candidate team fit.

Hiring Manager's Role:

The Hiring Manager's role matters because they:

1. Define job requirements: Clearly outline the skills and qualifications needed.
2. Evaluate candidate fit: Assess the technical skills and cultural alignment of all candidates.
3. Make the final decision: Choose the best candidate for the role.
4. Ensure team alignment: Select a candidate who will contribute to the team's success and growth.

Their expertise and understanding of the team's needs make their input crucial in the hiring process.

TIP: Speak their language. Demonstrate to them that you understand their team's pain points and how your candidate can address them.

Recruiter's Role

The Recruiter's role matters because they:

1. Source top talent: Identify and attract high-quality candidates.
2. Streamline the process: Manage logistics, scheduling, and communication.
3. Initial screening: Assess candidate qualifications and fit.
4. Build candidate relationships: Foster positive interactions and represent the company.

Their network helps efficiently find the best candidates, saving time and resources.

TIP: Strong recruiters don't just "submit resumes"; they advise, interpret, and deliver outcomes.

Department Head's Role

The Department Head's role matters because they:

1. Validate job requirements: Ensure the role aligns with departmental needs.
2. Provide context: Share insights into team dynamics and challenges that will help inform the discussion.
3. Assess candidate fit: Evaluate candidates' skills and experience.
4. Approve hiring decisions: Confirm the selected candidate meets departmental standards.

Their understanding of the departmental goals helps ensure the new hire will contribute to the team's success.

TIP: Engage them early in intake discussions when possible; they help shape hiring direction.

HR Representative's Role

An HR (Human Resources) Representative is a professional responsible for managing and supporting various aspects of an organization's workforce, including recruitment and hiring, employee relations and communications, benefits and compensation, compliance with labor laws and regulations, and training and development.

The HR Representative's role matters because they:

1. Ensure compliance: Verify adherence to company policies and procedures.
2. Maintain fairness: Oversee the hiring process to prevent bias and ensure equal opportunities for all candidates.
3. Provide guidance: Offer expertise on employment laws and regulations.
4. Support hiring managers: Assist with paperwork, onboarding, and other administrative tasks as needed.

They serve as liaisons between employees, management, and the organization, ensuring HR policies and procedures are followed. Their involvement helps ensure a smooth, compliant, and equitable hiring process.

TIP: Stay in close sync, especially if you are managing remote or international hiring logistics.

Team Members' Role

Team Members' role matters because they:

1. Provide insights: Share knowledge about the team's dynamics and challenges.
2. Assess candidate fit: Evaluate how well a candidate will integrate with the team.
3. Contribute to team success: Help ensure the new hire will support team goals.
4. Participate in interviews: Offer perspectives on candidates' skills and fit.

Their input helps ensure the new hire will be a good fit for the team and contribute to its success.

TIP: Candidates may be nervous around peers, so prepare them and gather informal feedback after the interview.

HOW TO READ AND USE IT ORGANIZATION CHARTS DURING SOURCING AND INTAKE DISCUSSIONS.

Before we discuss how to use IT organization charts, it is essential to understand their meaning. So, what is, or what are IT organization charts?

Firstly, IT Organization refers to the department or team within a company responsible for managing and supporting information technology systems, infrastructures, and services. This includes IT staff and personnel responsible for IT operations, support, and development. Technology infrastructure: Hardware, software, networks, and other technical assets. IT services and support; Help desk, troubleshooting, maintenance, and other support functions.

IT Organization Charts.

IT Organization Charts are visual representations of an organization's Information Technology (IT) department's structure, hierarchy, and relationships.

These charts typically display:

1. Roles and positions: Job titles, names, and departments within the IT organization.
2. Reporting relationships: Lines and connections showing who reports to whom.
3. Hierarchy and levels: Management layers, teams, and departments.

IT Organization Charts generally help to:

1. Clarify roles and responsibilities
2. Identify key stakeholders and decision-makers
3. Understand communication channels and workflows
4. Plan and manage IT projects and initiatives

By providing a clear and concise visual representation of the IT organization's structure, IT organizational charts facilitate effective communication, collaboration, and decision-making.

WHY USE ORG CHARTS IN RECRUITMENT?

Purpose	Benefit to Recruiter
Identify key decision-makers	Know who to build relationships with during intake and feedback cycles.
Understand team dynamics	Anticipate culture fit and collaboration needs.
Customize role pitches	Show how the candidate supports team goals, not just skills.
Clarify structure during intake calls.	Ask more thoughtful questions and reduce back-and-forth.
Map gaps in team structure	Identify upcoming or hidden needs beyond current openings.

Therefore, we will proceed directly to the use of the IT Organization Chart. Let's proceed.

Mastoring IT Org Charts: A Comprehensive Guide to Sourcing and Intake Discussions

Understanding and effectively utilizing IT organizational charts (org charts) is a crucial skill for anyone involved in sourcing and intake discussions. These visual representations of an organization's structure and hierarchy can provide invaluable insights, helping you navigate complex IT landscapes, identify key stakeholders, and make informed decisions. In this guide, we will explore the art of reading and using IT organization charts to elevate your sourcing and intake discussions.

Understanding the Basics of IT Org Charts

Before diving into the ways of using IT org charts, it's essential to know and understand the basics of it. Of course, you can't understand the subsequent parts perfectly well if you do not understand the basics. So, let's get right into it. An IT org chart typically consists of the following:

1. Boxes or nodes: These would represent individual roles or positions within the organization.
2. Lines and connections: Indicating reporting relationships, hierarchies, and communication channels.
3. Labels and annotations: Providing additional context, such as job titles, names, and departmental affiliations.

How to Read IT Org Charts:

To extract meaningful information from IT org charts, follow these steps:

1. Start at the top: Identify C-suite or senior tech leadership (CTO, CIO, VP of engineering)
2. Identify the organization's structure: Determine the overall hierarchy, including departments, teams, and reporting lines.
3. Locate key stakeholders: Find the names and titles of influential individuals, such as IT managers, directors, or CIOs.
4. Analyze departmental relationships: Examine how different departments interact and collaborate. Cross-functional roles, such as cross-functional product QA or DevOps, often work outside strict hierarchies.
5. Note any gaps or redundancies: Identify potential areas for improvement or optimization. Is there a missing function? (eg. Security)? This might reveal future hiring needs.

6. Confirm with the client: Ask: "Has this chart changed recently?" or "Where will the new hire sit in this structure?"

Using IT Org Charts in Sourcing Discussions:

When engaging in sourcing discussions, IT org charts can be a powerful tool:

1. Identify decision-makers: Determine who has the authority to make purchasing decisions or approve projects. And who has the overall power.
2. Understand technical requirements: Analyze the organization's technical infrastructure and needs.
3. Develop targeted solutions: Tailor your proposals to address specific pain points and challenges that are relevant to your audience.
4. Build relationships: Establish connections with key stakeholders and influencers.

Leveraging IT Org Charts in Intake Discussions

During intake discussions, IT org charts can help you:

1. Clarify project scope and objectives: Ensure you understand the project's goals, timelines, and stakeholders. Once all this has been understood, you already have an edge, as you will know what to expect, how the company is performing, and what others are doing.
2. Identify potential roadblocks roadblocks: Roadblocks can also be referred to as challenges or problems. To leverage IT Organization charts, you have to anticipate challenges and develop strategies to overcome them.
3. Develop effective communication plans: Establish clear channels of communication with key stakeholders to ensure open and transparent communication. Please make sure they are direct and also respect the offices of the stakeholders.
4. Create realistic project timelines: Estimate project durations and milestones based on the organization's structure and resources.

Advanced Techniques for Reading and Using IT Org Charts

After learning the basics of using an IT Organization Chart, it's only right you build on that knowledge. So, To take your skills of reading IT Organization Charts to the next level, do the following:

1. Look for patterns and trends: Identify commonalities in job titles, departmental structures, or reporting lines.
2. Analyze org chart changes: Study changes in the organization's structure over time to understand shifts in priorities or strategies.
3. Use org charts to identify opportunities: Recognize areas where your solutions or services can address specific pain points or challenges.
4. Integrate organization charts with other tools: Combine org charts with other data sources, such as company reports or industry research, to gain a more comprehensive understanding.

Best Practices for Working with IT Org Charts

When using an IT Organization chart, there are some things you do to ensure that the use of the chart remains seamless and smooth. They help you understand it even more and make it worthwhile even for people who do not know how to use an IT org chart. To maximize the value of IT org charts, you should:

1. Verify accuracy: Ensure the org chart is up-to-date and accurate.
2. Use multiple sources: Consult multiple sources to gain a more comprehensive understanding of the topic.
3. Analyze the big picture: Consider the organization's overall strategy and goals.
4. Keep org charts organized: Store and maintain org charts in a centralized location for easy access.

By mastering the art of reading and using IT org charts, you'll be better equipped to navigate complex IT landscapes, build strong relationships with key stakeholders, and drive successful sourcing and intake discussions.

Org chart questions to ask during intake.

- "Who will this role report to, and who will they work with most frequently?"

- "Are they part of an existing team, or will this be a net-new function?"
- "Will this person interface with Product, DevOps, or security teams?"
- "Where does this function sit in your company, under Engineering, Business, or IT Ops?"

ADVANCED USE: COMPARING ORG CHARTS ACROSS CLIENTS.

Over time, you'll notice patterns:

Company Type	Common Structure
SaaS Startups	Flat organs, lots of hybrid roles, direct report to CTO
Healthcare Enterprises	Layered structures, heavy QA, multiple compliance checkpoints
Product firms (Agile)	Squad-based organs with cross-functional pods (QA, DEV, PM)
FinTech	High presence of security, risk, GRC, and audit oversight

WHERE DIFFERENT ROLES SIT WITHIN TECHNICAL AND BUSINESS TEAMS.

In today's fast-paced and increasingly digital business landscape, the lines between technical and business teams are becoming increasingly blurred. IT roles are no longer confined to traditional technical departments; instead, they are integrated into various aspects of the organization. In this note, we will examine the roles of different IT professionals within technical and business teams, as well as their contributions to the organization's overall success.

Technical Teams

Technical teams are typically responsible for designing, developing, testing, and implementing technology solutions. Within these teams, you can find a variety of IT roles, including:

1. Software Developers/Engineers: Responsible for designing, developing, and testing software applications.
2. DevOps Engineers: Focus on ensuring the smooth operation of software systems, from development to deployment.
3. Quality Assurance (QA) Engineers: Test and validate software applications to ensure they meet requirements and are free from defects.

4. Data Scientists/Engineers: Analyze and interpret complex data to inform business decisions and drive strategy.
5. Cybersecurity Specialists: Protect the organization's technology assets from cyber threats and vulnerabilities.

Business Teams

Business teams, on the other hand, are focused on driving business strategy, operations, and growth. While they may not be directly responsible for technical implementation, they rely heavily on IT to inform their decisions and drive business outcomes. IT roles within business teams include:

1. Business Analysts: Work with stakeholders to identify business needs and develop solutions to address them.
2. Product Managers: Oversee the development and launch of products, working closely with technical teams to ensure successful delivery and implementation.
3. Project Managers: Coordinate and manage projects, ensuring they are completed on time, within budget, and to the required quality standards.
4. Business Intelligence Analysts: Analyze data to inform business decisions and increase their strategy.
5. Digital Transformation Specialists: Assist organizations in navigating digital change, leveraging technology to drive business growth and innovation.

Hybrid Roles

As technology becomes increasingly integral to business operations, hybrid roles are emerging that combine technical and business skills. Examples include:

1. Technical Product Managers: Combine technical expertise with business knowledge to increase product development and launch.
2. Business Technologist*: Work at the intersection of business and technology, using technical skills as a way to increase business outcomes.
3. Digital Business Analysts: Analyze business needs and develop solutions that use technology to drive growth and innovation.

Where IT Roles Sit within Technical and Business Teams

To illustrate where different IT roles sit within technical and business teams, you have to know that both business and technical roles depend on IT, as they are a way to make things easier. See the following examples of IT roles:

1. Software Developers: Typically report to a technical lead or manager within a technical team.
2. Business Analysts: May report to a business team lead or manager but often work closely with technical teams.
3. Product Managers: Typically report to a business team lead or manager but work closely with technical teams to drive product development.
4. Data Scientists/Engineers: May report to a technical lead or manager within a technical team but often work closely with business teams to inform business decisions.
5. Cybersecurity Specialists: Typically report to a technical lead or manager within a technical team but may work closely with business teams to ensure compliance and risk management.

In conclusion, IT roles are no longer confined to traditional technical departments but are instead integrated into various aspects of the organization. By understanding the placement of different IT roles within technical and business teams, organizations can better leverage their technical expertise to drive business growth and innovation. Whether you're a technical professional or a business leader, recognizing the dynamics of IT roles within technical and business teams is essential for success in today's digital landscape.

WHY ROLE PLACEMENT MATTERS:

When you understand where a role fits,

- You screen better: You'll know whether they need deep coding or stakeholder experiences.
- You ask sharper intake questions, e.g., "Will they be customer-facing?" or "Who approves scope changes?'
- You close stronger: Candidates gain clarity on whom they'll work with and report to.

Fitcheck: Role and placement

Role	Sits with	Key screening focus
DevOps Engineer	Infra/Tech	IaC, automation, AWS tools
BSA- Healthcare	Business/IT Hybrid	EMR/EHR, Stakeholder communication, HIPAA
QA Analyst	QA/engineering	Test cases, automation tools, defect lifecycle
Cloud Architect	Tech architecture	Multi-cloud experience, scaling, and cost optimization.
Customer support rep	Business support	CRM familiarity, escalation handling

BUILDING CONFIDENCE IN MAPPING CANDIDATES TO TEAM STRUCTURES AND CLIENT NEEDS.

As a recruiter or hiring manager, one of the most critical tasks is to identify and map candidates to team structures and client needs. This process requires a deep understanding of the organization's goals, team dynamics, and the skills needed to achieve success. However, building confidence in this process can be challenging, especially when dealing with complex team structures and diverse client needs. In this note, we will be exploring the strategies and best practices for building confidence in mapping candidates to team structures and client needs.

Understanding the Team Structure

Before mapping candidates to team structures, it is essential to have a clear understanding of the organization's team dynamics. This includes:

1. Departmental goals and objectives: Understanding the department's mission, vision, and key performance indicators (KPIs).
2. Team roles and responsibilities: Identifying the key roles and responsibilities within the team, including the skills and expertise required.
3. Reporting structures and hierarchies: Understanding the organizational chart, including reporting lines and decision-making processes.

4. Communication channels and workflows: Identifying the key communication channels and workflows within the team.

Identifying Client Needs

To map candidates to client needs, you must have a deep understanding of the client's requirements and expectations. This includes:

1. Client goals and objectives: Understanding the client's mission, vision, and KPIs.
2. Client pain points and challenges: Identifying the client's key challenges and pain points.
3. Client culture and values: Understanding the client's culture, values, and work environment.
4. Client expectations and requirements: Identifying the client's expectations and requirements for the role.

Mapping Candidates to Team Structures and Client Needs

With a clear understanding of the team structure and client needs, you can begin mapping candidates to these requirements. This involves:

1. Identifying key skills and qualifications: Determining the essential skills and qualifications required for the role.
2. Assessing candidate fit: Evaluating the candidate's skills, experience, and fit for the role and team.
3. Evaluating candidate strengths and weaknesses: Identifying the candidate's strengths and weaknesses and how they align with the team's and client's needs.
4. Considering team dynamics and culture: Assessing the candidate's fit with the team culture and dynamics.

STEP BY STEP: How to map candidates effectively

1. Understand the team structure; ASK: 'Who is the hiring manager, and where do they sit?' 'Who will the candidate work with directly?' 'What tools, methodologies, or workflows does the team use?'
2. Clarify the business need. Examples: 'We need to migrate from on-prem to AWS', 'We are rolling out a new EHR system', and 'Our QA team lacks automation expertise."

3. Assess Candidate Fit: Go beyond the resume: 'Does their experience align with the team's structure and goals?' 'Do they complement the team's skill mix?' 'Will they thrive in a matrixed or fast-paced setup?'
4. Map and Present Confidently: Use charts, visuals, or matrices to explain to your client: 'Why this candidate fits the gap', 'How they plug into the team hierarchy', and' What outcomes they're positioned to drive.'

Strategies for Building Confidence

To build confidence in mapping candidates to team structures and client needs, consider the following strategies:

1. Develop a deep understanding of the organization: Take the time to learn about the organization's goals, team dynamics, and culture.
2. Build relationships with hiring managers and clients: Establish strong relationships with hiring managers and clients to gain a deeper understanding of their needs and expectations.
3. Use data and analytics: Leverage data and analytics to inform your decision-making and identify trends and patterns.
4. Develop a robust candidate assessment process: Create a comprehensive candidate assessment process that evaluates skills, experience, and fit.
5. Continuously evaluate and improve: Regularly assess and refine your mapping process to ensure it remains effective and efficient.

Best Practices for Mapping Candidates

To ensure accurate and effective mapping, consider the following best practices:

1. Use a competency-based approach: Focus on the key competencies required for the role and evaluate candidates against these competencies.
2. Use a combination of assessment tools: Leverage a range of assessment tools, including interviews, skills tests, and reference checks.
3. Involve multiple stakeholders: Engage with various stakeholders, including hiring managers, clients, and team members, to gain a comprehensive understanding of the requirements.
4. Consider the candidate's career goals and aspirations: Evaluate the candidate's career goals and aspirations to ensure alignment with the role and organization.

5. Provide feedback and coaching: Offer feedback and coaching to candidates to help them develop and grow.

Overcoming Common Challenges

When mapping candidates to team structures and client needs, you may encounter common challenges, such as:

1. Limited candidate pool: Strategies for expanding the candidate pool include leveraging social media, employee referrals, and targeted advertising.
2. Insufficient information about the client or team: Strategies for gathering more details include conducting stakeholder interviews, reviewing job descriptions, and analyzing industry trends.
3. Difficulty evaluating candidate fit: Strategies for assessing candidate fit include using behavioral interviews, skills tests, and reference checks.

Building confidence in mapping candidates to team structures and client needs requires a deep understanding of the organization, team dynamics, and client requirements. By developing a robust candidate assessment process, leveraging data and analytics, and continuously evaluating and improving your mapping process, you can ensure accurate and effective mapping. By following the strategies and best practices outlined in this note, you'll be well on your way to building confidence in mapping candidates to team structures and client needs.

The Power of Understanding Client Team Structures

Knowing how your client's teams are structured can give you a significant advantage. When you are aware of the specific roles and responsibilities within a team, such as where a Cloud Engineer fits in a DevOps team or who a QA Analyst reports to, you can:

- Refine your screening process: Tailor your candidate search to meet the specific needs of the team and role.
- Improve communication: Clearly articulate your understanding of the team's structure and needs to build trust with your clients.

Increase client trust by Demonstrating your expertise and attention to detail, which leads to stronger relationships and increased confidence in your services.

By taking the time to understand your client's team structures, you can deliver more effective solutions and build stronger partnerships.

Organizational Models in IT

There are several organizational models commonly used in IT:

1. Agile/Scrum Teams

Cross-functional and project-based teams. They use iterative and incremental approaches to deliver projects.

Their Roles include:

- Product Owner
- Scrum Master
- Developers
- QA
- UX

2. Enterprise IT Structures

It is common in organizations with complex IT needs. It is more layered with specific departments, such as:

- Infrastructure
- Application Development
- Support
- Security
- Often includes roles like:
- Business Analysts
- Solution Architects
- Project Managers

3. Matrix Structures

- Employees report to both functional and project managers
- Common in consulting firms or large tech corporations

These organizational models can impact how teams collaborate, communicate, and deliver projects. Understanding these structures can help IT professionals navigate their organizations and work effectively with stakeholders.

Visual layouts showing hierarchy in an agile squad, highlighting how developers, testers, and leads collaborate under a scrum master and product owner.

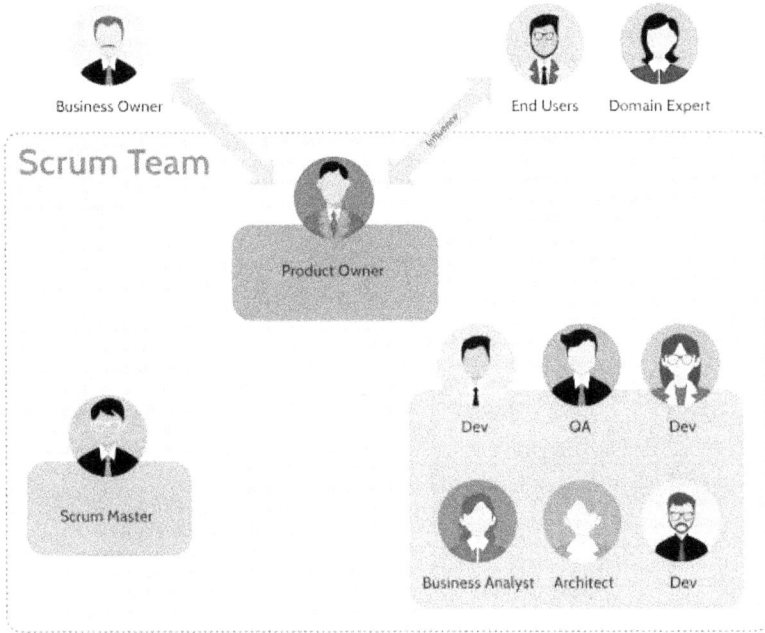

Key Agile Scrum Team Model Roles

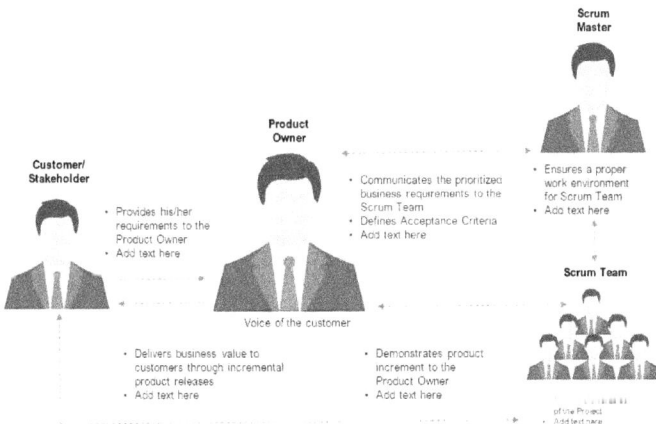

SAMPLE ENTERPRISE IT ORGANIZATION CHART.

ROLE-NEED MAPPING MATRIX.

JOB TITLE	REPORTS TO	DEPARTMENT	BUSINESS NEED.
1 QA Analyst	QA lead/ Test manager	QA/ Engineering	Ensures product quality.
2 Cloud Engineer	DevOps manager	Infrastructure	Enables scalable environments.
3 BSA (Healthcare)	Product Manager	Business/ IT	Captures and translates needs.
4 Software Developer	IT manager	Develops software applications	Meet technology needs and improve efficiency.
5 Customer Service Representative	Customer Service department	Handles customer inquiries and issues	Improves customer attention and retention.

Using Organization Charts in Recruitment

Understanding the Organization

During intake calls, ask questions like:

- "Where does this role fit in your organization?"
- "Who will they interact with daily?"

This helps you understand the team's structure and how it works. This way, you can relate with the candidate and let them know about it before closing the deal.

Evaluating Fit

Use organization chart knowledge to assess:

- Cultural fit: How will the candidate interact with the team?
- Technical fit: How will the candidate contribute to the team's goals?

This is also an essential aspect, as you would need to know a few other things beyond just the company's structure. You should let the client know the company's culture.

Closing Candidates

When closing candidates, explain:

- Team dynamics: Describe the team's culture and work style.
- Role within the team: Highlight how the candidate's skills and experience align with the team's needs.

This helps build confidence and reduces the number of dropped offers.

CASE EXAMPLE: Let's Break It Down

Imagine working with a company that develops apps for mobile phones. They need someone to help the different teams work more effectively together.

The Problem:

The company requested a "QA Lead," which typically refers to someone who tests apps to identify errors. However, they wanted someone to act as a "team captain" who helps the developers, testers, and operations teams work together smoothly.

The Solution:

The Recruiter asked for an organization chart, which is like a map of who is who in the company. This helped them understand who the new person would work with and who they would report to.

The Recruiter soon found the right person for the job, someone who could help the teams work together better.

Asking questions and getting a clear understanding of what the company needs is super important. In the company's culture, a QA lead performs a different task than in other companies, and the Recruiter would have made a mistake in recruiting the wrong candidate if they had not consulted the organization chart. Employing the wrong candidate would have led to "offer dropouts," which means the candidate leaving just after a few days because they do not understand the system or achieve below-standard results.

NOTE: This example is scenario-based for training and learning purposes, designed to help us better understand the scope. The real scope can be much deeper and more precise.

Self Assessment 3

1. What are the key components of an Agile team structure?
2. How is an Enterprise IT structure different from an Agile team?
3. Why is understanding reporting relationships meaningful in recruitment?
4. What is a matrix organization?
5. How can knowing where a role fits help reduce offer dropouts?

Reflective Prompt: Think of a recent job role you sourced for. Where does that role fit in the client's organization? Did you ask or assume? What could you have done differently to understand the structure better?- Reading and Understanding IT Org Charts.

CHAPTER FOUR

HIRING MANAGER AND PROJECT MANAGER PAIN POINTS

Objectives of this learning:

- Understand the expectations and frustrations of hiring managers and project managers
- Identify the most common recruitment-related pain points on the client side
- Learn how to ask meaningful intake call questions that uncover real needs
- Build strategies for aligning recruitment efforts with project goals and team dynamics
- Develop trust-based, consultative relationships with clients

Who is a hiring manager?

A hiring manager is typically a person responsible for recruiting, interviewing, hiring, and supervising new hires in a company. A hiring manager works closely with Departmental heads, team leads, line managers, talent acquisition specialists, and others. Their goal is to identify the most suitable candidate for the job.

Qualities of a Good Hiring Manager

Before a person can be described as a good hiring manager, there are some qualities he should possess. These qualities would lead to the growth and development of the company, as they would facilitate the hiring of qualified candidates. Here are some key attributes:

1. Clear Communicator: Effectively conveys job expectations, company culture, and vision.
2. Fair and Unbiased: Makes hiring decisions based on merit, skills, and fit.
3. Good Listener: Actively listens to candidates and understands their needs and concerns.

4. Knowledgeable: Familiar with the job requirements, industry, and company.
5. Decisive: Makes timely and informed hiring decisions.
6. Empathetic: Understands candidate perspectives and creates a positive interview experience.
7. Organized: Manages the hiring process efficiently, ensuring timely follow-ups and effective communication.
8. Strategic Thinker: Aligns hiring decisions with company goals and plans.

By possessing these qualities, hiring managers can:

1. Attract top talent
2. Improve candidate experience
3. Make informed hiring decisions
4. Enhance team performance

Ultimately, good hiring managers play a crucial role in building high-performing teams.

Expectations and Frustrations of Hiring Managers and Project Managers

As a hiring manager or project manager, you have a lot on your plate. You're responsible for finding and hiring the best talent, managing projects, and ensuring everything runs smoothly. But what are your expectations and frustrations? Let's dive in.

Hiring Managers' Expectations:

Before or when hiring a candidate, a hiring manager has some expectations. Some of these include:

1. Finding the Perfect Candidate: You want someone who fits the job description, has the right skills, and can do the job well.
2. Streamlined Hiring Process: You expect a smooth and efficient hiring process with minimal delays and complications.
3. Quality Over Quantity: You would rather have a few highly qualified candidates than a bunch of unqualified ones.
4. Cultural Fit: You want someone who fits in with your company's culture and values.

5. Timely Hiring: You need to fill the position promptly so your team can return to work.

Hiring Managers' Frustrations:

1. Poor Candidate Quality: You are frustrated when candidates do not meet the job requirements or lack the necessary skills.
2. Lengthy Hiring Process: You are annoyed when the hiring process takes too long, causing delays and impacting your team's productivity.
3. Unrealistic Candidate Expectations: You are frustrated when candidates have unrealistic expectations about the job or company.
4. Lack of Communication: You are irritated when there is poor communication from recruiters or HR, leaving you in the dark about the hiring process. Communication is key in every process.
5. Budget Constraints: You are limited by budget constraints, making it difficult to attract top talent. The hiring process involves a significant amount of funds, and often, the hiring manager or the company may not have the necessary funds to cover these expenses.

PROJECT MANAGER.

Who is a project manager?

A project manager is a professional responsible for planning, organizing, and controlling all the resources to achieve specific goals and objectives in a company within a particular time, budget, and others. He is involved in project planning, team management, risk management, communication, and monitoring the progress of all ongoing activities.

Qualities of a Good Project Manager

A good project manager should possess some skills and qualities that would ease their work and make all their tasks smooth and efficient. Here are some key attributes that would classify one as a good project manager:

Leadership Qualities

A good project manager must possess strong leadership qualities, as they will help him be an effective leader. Some of these leadership qualities are:

1. Strong Communication: A good project manager should be able to convey project goals, expectations, and progress clearly.
2. Visionary: He must align project goals with organizational objectives.
3. Motivational: He must be able to inspire and motivate team members. A good project manager should not be the reason for a loss of motivation among team members. He should be able to keep them motivated at all times.

Organizational Skills

Apart from leadership qualities, a good project manager should have organizational skills as they would help him properly plan and deliver his tasks. Some of these organizational skills include:

1. Planning: A good project manager must be able to develop detailed project plans and timelines that accurately reflect the project's scope and requirements.
2. Prioritization: He must be able to prioritize tasks and allocate resources effectively.
3. Risk Management: Identifies, assesses, and mitigates risks. A good project manager should be able to take calculated risks and also be accountable for them in case they go wrong.

Interpersonal Skills

A good project manager should also possess some personal skills. How he relates to people or things matters a lot. Some of these skills are:

1. Collaboration: He should be able to foster teamwork and collaboration among stakeholders and every other person in the company, regardless of their level or ranking.
2. Emotional Intelligence: Understands and manages emotions, conflicts, and stakeholder expectations. He should be able to resolve disputes before they escalate, prevent tension, and use his words carefully to avoid hurting the egos of any candidate, etc.
3. Adaptability: He should be able to adapt to changing project requirements and stakeholder needs.

Technical Skills

A good project manager must also possess strong technical skills; they should have the necessary knowledge and expertise required to perform a particular job or task. Some of these skills are:

1. Project Management Methodologies: Familiarity with Agile, Waterfall, or hybrid approaches. All of these approaches are essential for being a good project manager, and he must have studied and understood them.
2. Tools and Software: Proficiency in project management tools, such as Asana, Trello, or MS Project.
3. Analytical: He should be able to analyze data to inform project decisions.

Other Essential Qualities of a good project manager include:

1. Accountability: Takes ownership of project outcomes. As previously stated, a good project manager should be accountable for the risks they take, regardless of the outcomes. He should not be a liar.
2. Flexibility: Adapts to changing circumstances. Things change every day with new developments in the world. He should be able to adapt quickly and not be left out.
3. Continuous Learning: Stays up to date with industry trends and best practices. As new developments arise, he should be able to learn. If there are new technological devices or developments, he should be able to keep learning how to use them.

By possessing these qualities, project managers can deliver successful projects, build strong teams, and drive business results.

Project Managers' Expectations.

After being a good project manager, you now expect more from the candidates. Some of your expectations should include:

1. Clear Project Goals: You want clear project goals, objectives, and timelines.
2. Effective Communication: You expect open and transparent communication with your team, stakeholders, and clients.

3. Timely Deliverables: You need team members to deliver high-quality work on time.
4. Collaborative Team: You want a team that works well together, collaborates effectively, and supports one another.
5. Flexibility: You expect to be able to adapt to changing project requirements and priorities.

Project Managers' Frustrations.

These are things that would likely stress you out as a project manager, even after putting in your best effort. Some of them are:

1. Unclear Project Goals: You are frustrated when project goals are unclear or constantly changing.
2. Poor Communication: You become annoyed when there is poor communication within the team or with stakeholders, as this slows down your efforts and leaves you feeling lost and uninformed about what is happening.
3. Delays and Missed Deadlines: You are frustrated when team members miss deadlines or deliver low-quality work, as this can annoy the client and lead to them returning the job or not paying in full, resulting in a loss to the company.
4. Lack of Resources: You are limited by lack of resources, including budget, personnel, or equipment.
5. Unrealistic Expectations: You are frustrated when stakeholders or clients have unrealistic expectations about the project's scope, timeline, or budget.

RECRUITMENT-RELATED PAIN POINTS ON THE CLIENT's SIDE.

Now, we have spoken about the project manager and the hiring manager. How about as a client?

As a client, you have likely experienced frustration with the recruitment process. Here are some common pain points:

1. Difficulty Finding Qualified Candidates

- Lack of relevant experience: Candidates may not have the specific skills or experience required for the role.

- Insufficient talent pool: The candidate pool may be limited, making it challenging to find the right person.
- Poor candidate quality: Candidates may not meet the job requirements or lack the necessary skills.

2. Lengthy Hiring Process

- Time-consuming interviews: Multiple rounds of interviews can delay the hiring process.
- Slow candidate screening: Delays in screening and shortlisting candidates can prolong the hiring process.
- Unresponsive recruiters: Lack of communication from recruiters can leave clients wondering about the status of their job postings.

3. High Recruitment Costs

- Cost per hire: Recruitment agencies may charge high fees, increasing the cost per hire.
- Advertising expenses: Job postings and advertising can be costly, especially for hard-to-fill positions.
- Training and onboarding: Investing time and resources in training new hires can be expensive.

4. Poor Candidate Fit

- Cultural mismatch: Candidates may not align with the company's culture or values.
- Skills mismatch: Candidates may not have the required skills or experience.
- Personality clashes: Candidates may not get along with existing team members.

5. Lack of Transparency and Communication

- Poor communication from recruiters: Clients may not receive regular updates on the hiring process.
- Lack of feedback: Candidates may not receive feedback on their application or interview performance.
- Unclear expectations: Clients may not have a clear understanding of the recruitment process or timelines.

6. Difficulty Attracting Top Talent

- Competition from other companies: Other companies may be offering more attractive salaries, benefits, or opportunities.
- Lack of employer branding: Companies may lack a strong employer brand, making it difficult to attract top talent.
- Insufficient benefits and perks: Companies may not offer competitive benefits and perks to attract top candidates.

7. Inefficient Recruitment Process

- Manual processes: Manual processes, such as screening resumes and scheduling interviews, can be time-consuming and labor-intensive.
- Lack of technology: Companies may not be using recruitment technology, such as applicant tracking systems (ATS), to streamline the hiring process.
- Inefficient workflows: Workflows may not be optimized, leading to delays and inefficiencies.

By understanding these pain points, recruitment agencies and companies can collaborate to enhance the hiring process, minimize costs, and identify top talent.

Asking Meaningful Intake Call Questions

Before learning how to ask meaningful intake questions, it is essential to understand what they entail. So, what are meaningful intake questions?

Meaningful intake questions are specific, open-ended questions asked during the initial consultation or intake call to understand the client's needs and goals, gather detailed information about job requirements, uncover the client's expectations and priorities, and identify potential challenges and areas of concern.

All these help ensure a clear understanding of the client's requirements, enabling better matching of candidates and more effective solutions.

So, On that note, when conducting an intake call, your goal is to uncover the client's real needs and understand their requirements. Here are some tips and questions to help you achieve this:

Before the Call:

1. Research the client: Review the client's website, industry, and company culture to gain a comprehensive understanding of their business. Do not unthinkingly enter into it without knowing anything about it.
2. Review the job description: Familiarize yourself with the job description and requirements.

During the Call:

Introduction and Context

1. Confirm the job requirements: Verify the job title, responsibilities, and requirements.
2. Understand the client's goals: Ask the client about their goals and objectives for the role.

Uncovering Real Needs

These include the actual questions to be asked and why they should be asked.

1. QUESTION: What are the biggest challenges facing the team or department right now?

- This question helps you understand the team's pain points and what the client is trying to achieve.

2. QUESTION: Can you describe the ideal candidate's work style and personality?

- This question helps you understand the client's expectations and determine the type of candidate who would be a good fit for the team.

3. QUESTION: What are the most essential skills and qualifications for this role?

- This question helps you understand the client's priorities and what they value most in a candidate.

4. QUESTION: How will success be measured in this role?

- This question helps you understand the client's expectations and how the candidate's performance will be evaluated.

5. QUESTION: Are there any specific pain points or challenges you've experienced with previous hires?

- This question helps you understand what the client is trying to avoid in future hires.

Questions on Cultural Fit and Team Dynamics

1. QUESTION: Can you describe the company culture and values?

- This question helps you understand the client's expectations and what type of candidate would fit in with the company culture to prevent employing candidates who can't survive or work with the company's values.

2. QUESTION: How does the team collaborate and communicate?

- This question helps you understand the team's dynamics and what type of candidate would thrive in that environment.

Questions on Next Steps and Expectations

1. QUESTION: What is the timeline for filling this position?

- This question helps you understand the client's expectations and prioritize your efforts accordingly.

2. QUESTION: What are the next steps in the hiring process?

- This question helps you understand the client's process and what to expect.

After the Call

1. Summarize the key points: Recap the client's needs, expectations, and requirements.
2. Clarify any doubts: Ask follow-up questions to ensure you understand the client's needs.
3. Document the conversation: Take detailed notes and record the conversation for future reference.

By asking these meaningful intake call questions, you can uncover the client's actual needs, understand their expectations, and deliver a more effective service.

ALIGNING RECRUITMENT EFFORTS WITH PROJECT GOALS AND TEAM DYNAMICS.

To develop effective recruitment strategies, it is crucial to comprehend the project's objectives, team dynamics, and the specific skills necessary to achieve success. Here is a step-by-step, carefully compiled guide to help you align your recruitment efforts:

Understanding Project Goals

1. Review project objectives: Familiarize yourself with the project's goals, timelines, and deliverables.
2. Identify Key Performance Indicators (KPIs): Understand how success will be measured and what metrics are crucial to the project's success so time is not wasted on activities that may not contribute to the company's overall success.
3. Determine required skills and expertise: Identify the skills, knowledge, and experience necessary to achieve the project's objectives.

Understanding Team Dynamics

1. Team structure and roles: Understand the team's composition, roles, and responsibilities. Understand the entire structure. Understand the company's hierarchy system and the people in charge of some aspects to prevent future conflicts or confusion.
2. Communication styles: Familiarize yourself with the team's communication styles, preferences, and collaboration tools.
3. Team culture and values: Understand the team's culture, values, and work environment.

Recruitment Strategy

1. Know and understand the recruitment goals: Align recruitment goals with project objectives, focusing on finding candidates with the required skills and expertise.
2. Develop a candidate profile: Create a profile outlining the ideal candidate's skills, experience, and qualities.
3. Sourcing strategies: Identify effective sourcing channels, such as job boards, social media, or employee referrals.

4. Interview process: Develop an interview process that assesses candidates' technical skills, teamwork abilities, and problem-solving capabilities. To make things more understandable, the interview process should be comprehensive, focusing on all aspects of work energy. A candidate may have the right qualifications but may also have the wrong attitude towards work.
5. Assessment and Evaluation: Utilize relevant assessment and evaluation tools to ensure candidates meet the project's requirements.

Collaboration with Hiring Managers and Stakeholders.

1. Regular communication: Maintain open communication with hiring managers and stakeholders to ensure alignment and address concerns.
2. Feedback and iteration: Solicit feedback from hiring managers and stakeholders to refine the recruitment process and improve candidate quality.
3. Stakeholder involvement: Involve stakeholders in the recruitment process to ensure their needs are met and they are invested in the project's success. Stakeholders should be involved, as they often exert influence behind the scenes. If they decide not to support a particular project, it may be disregarded by everyone.

Benefits of Alignment

1. Improved candidate fit: Candidates are more likely to succeed in the role and fit in with the team.
2. Increased project success: Recruitment efforts support the project's objectives, contributing to its success. When everyone works in alignment with the project goals and team dynamics, the success rate increases.
3. Enhanced team dynamics: New hires are more likely to integrate seamlessly into the team, promoting a positive and productive work environment.

By following these steps and aligning recruitment efforts with project goals and team dynamics, you'll be able to attract top talent, improve candidate fit, enhance project success, and foster a positive and productive team environment for everyone, both recruits and existing staff.

Developing Trust-Based, Consultative Relationships with Clients.

Before we head deeper into this, we need to define trust.

Trust is the belief in the reliability, truth, ability, or strength of something or something. It is built on integrity, competence, and empathy.

A trust-based consultative relationship means building strong relationships with clients by understanding their needs and goals, providing expert advice and solutions, and demonstrating transparency, reliability, and empathy. This fosters collaboration between you and the client, and this is what we will be discussing now.

Building strong, trust-based relationships with clients is crucial for long-term success. Here's a step-by-step guide to help you develop consultative relationships:

Understanding Client Needs

1. Active listening: Listen attentively to clients, asking clarifying questions to ensure understanding.
2. Needs assessment: Conduct thorough needs assessments to identify clients' pain points and goals.
3. Client profiling: Create client profiles to track their preferences, interests, and communication styles.

Building Trust

1. Transparency: Be open and honest in all interactions, providing clear explanations and updates.
2. Reliability: Deliver on promises, meeting deadlines and expectations.
3. Expertise: Demonstrate industry knowledge and expertise, providing valuable insights and advice.
4. Empathy: Demonstrate understanding and empathy by acknowledging clients' challenges and concerns.

Consultative Approach

A consultative approach involves working closely with clients to understand their needs, providing expert advice, and then developing tailored solutions that meet their specific goals and objectives.

A consultative approach is usually:

1. Solution-focused: Focus on finding solutions that meet client's needs rather than just pushing products or services.
2. Collaborative: Work collaboratively with clients to identify opportunities and challenges.
3. Proactive: Anticipate client's needs and provide proactive solutions and recommendations.
4. Customized solutions: Develop tailored solutions that address clients' unique needs and goals.

Communication

The client must be informed about any developments that affect them directly or indirectly when there is a bridge in communication. Things begin to go wrong. Ways to achieve effective communication include:

1. Regular updates: Provide regular updates on progress to ensure clients are informed and engaged.
2. Clear communication: Use clear, concise language, avoiding jargon and technical terms unless necessary, as they could confuse clients who have no prior knowledge of them before meeting with you.
3. Responsive: Respond promptly to clients' queries and concerns, demonstrating attentiveness and care.

Long-Term Relationship Building.

It is also essential to build long-term relationships with your clients, as they may refer you to others in the future and be glad to work with you again. Ways to build a long-term relationship with clients include:

1. Consistency: Deliver high-quality service consistently, building trust and credibility. You can't be on and off with a client and expect a long-term relationship with them. When they are continuously disappointed, they will likely not want to work with you again, as that trust has been broken.
2. Continuous improvement: Continuously improve knowledge, skills, and processes to meet evolving client needs. Please demonstrate that

you are willing to learn and improve to ensure they receive what they want.

3. Feedback: Solicit feedback from clients at all times, using it to refine services and improve relationships. Do not wait until the job is almost done before getting feedback. If the client is not impressed, you will have to start all over again, which will waste time and displease the client. So, ask for feedback every step of the way.

Benefits of Trust-Based Relationships.

After building a trust-based relationship with your client, you should expect the following benefits:

1. Increased client loyalty: Clients are more likely to remain loyal and continue working with you.
2. Improved client satisfaction: Clients are more satisfied with services, leading to positive word of mouth and referrals.
3. Competitive advantage: Trust-based relationships differentiate you from competitors, making it harder for clients to switch because they are already sure you wouldn't disappoint them.

By following these steps and adopting a consultative approach, you can build strong, trust-based trust-based relationships with clients, driving long-term long-term success and growth.

Understanding Key Stakeholders

By recognizing hiring managers and project managers as key clients, you can tailor your approach to their needs and priorities, develop a deeper understanding of their pain points and challenges, provide more effective solutions and communication, and Build stronger, more valuable relationships. This enables you to deliver more targeted and impactful services, resulting in better outcomes for all stakeholders involved.

REAL QUOTES FROM HIRING MANAGERS.

Hiring managers often have unique perspectives and pain points when it comes to the recruitment process. Here are some quotes, along with explanations of why they made such quotes:

1. "I don't need 10 resumes. I need two who fit."

This quote highlights the hiring manager's desire for quality over quantity. They are looking for candidates who closely match the job requirements rather than reviewing numerous resumes that may not be relevant to the position. This is common during job applications or recruitments; there could be a lot of applications, ns but there is no perfect candidate for the job. This approach saves time and increases the chances of finding the right candidate.

2. "Stop sending me what the resume says. Tell me what the candidate can do."

Hiring managers often receive numerous resumes with similar skills and qualifications. This quote suggests that they want recruiters to go beyond just summarizing the candidate's resume. Instead, they want to know what the candidate can deliver, their strengths, and how they will contribute to the organization. It is also not a new phenomenon that some candidates inflate their resumes with false qualifications to secure the job, so they want to know if the candidates can perform with or without the qualifications listed on their resumes.

3. "If your candidate drops out, who's your backup?"

This quote demonstrates the hiring manager's concern about the reliability of the recruitment process. They want to know that the Recruiter has a plan in place in case the selected candidate does not work out. This indicates that hiring managers value preparedness and proactive problem-solving skills.

4. "I hate explaining my tech stack in every intake call. It wastes my time."

This quote highlights the frustration hiring managers experience when having to explain technical details to recruiters repeatedly. They want recruiters to have a basic understanding of their technology stack, allowing them to focus on more critical aspects of the recruitment process instead of just learning new things when they should have focused on the main point. This can be very stressful for hiring managers.

5. "I need someone who can hit the ground running, not someone who needs extensive training."

Hiring managers often prioritize candidates who can quickly adapt to the role and start making a positive contribution to the organization's success. This quote suggests that they value candidates with relevant experience and skills that align with their needs. This is why they do not just depend on resumes. Finding the perfect candidate means that the candidate must have some level of experience, perhaps gained through internships or volunteering.

6. "Can you find someone who understands our company culture?"

This quote emphasizes the importance of cultural fit in the hiring process. Hiring managers want candidates who not only have the necessary skills but also align with the company's values, mission, and work environment.

7. "I don't want to keep going through generic cover letters. Make sure your candidates match theirs to our company."

Hiring managers often receive numerous applications with generic cover letters that do not demonstrate a genuine interest in the company. This quote suggests that they value candidates who take the time to research and understand the company's specific needs and goals rather than those who copy and paste their application letters.

8. "What's your process for ensuring candidate quality?"

This quote highlights the hiring manager's concern about the quality of candidates being presented. They want to know that the Recruiter has a robust process in place to assess candidate skills, experience, and fit for the role.

So, Why Do Hiring Managers Say These Things?

Hiring managers say these things because they:

1. Value efficiency and productivity in the recruitment process
2. Want recruiters to understand their specific needs and challenges
3. Prioritize quality over quantity in candidate selection
4. Need recruiters to be proactive and solution-focused
5. Appreciate recruiters who take the time to understand their company's culture and goals.

FLOW CHART DIAGRAM

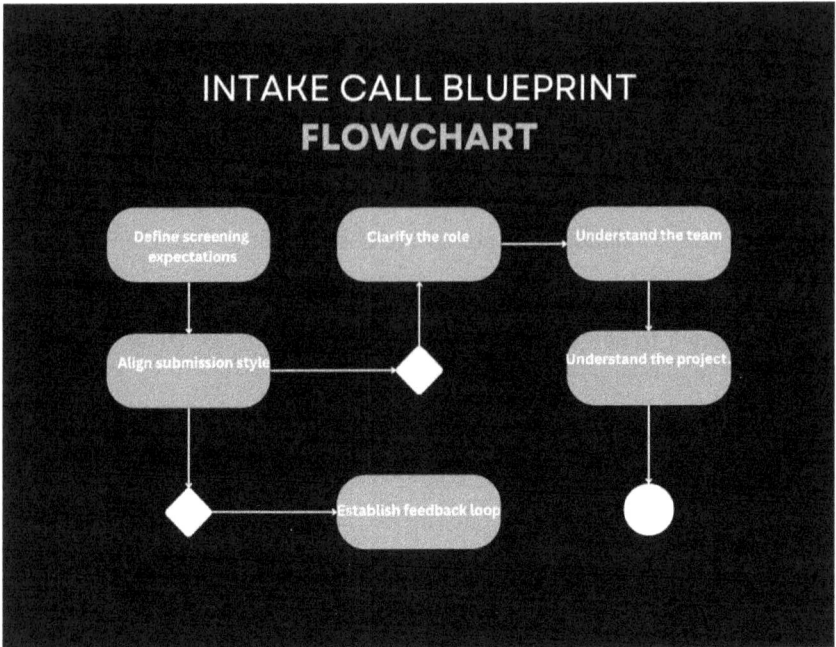

INTAKE CALL BLUEPRINT
FLOWCHART

1. Understand the project: You must understand the scope, urgency, and budget. You should know everything it entails.
2. Understand the team: You must understand the structure and the key collaborators. Apart from the other factors, you should understand the people.
3. Clarify the roles: The must-haves, nice-to-haves, and deal-breakers. Try to know what you must have or do clearly. Clearly explain the dos and don'ts.
4. Define screening expectations
5. Align submission style: Whether you are using Email, ATS, or pitch format, the submission style should align in some way.
6. Establish a feedback loop; Feedback is necessary. Always make sure you receive feedback and try to send yours as well.

CONDUCTING A HIGH-IMPACT INTAKE CALL.

A high-impact intake call is crucial in setting the tone for a successful recruitment process. Here is how to make the most out of it:

1. Ask about project goals, not just skills.

Instead of solely focusing on the technical skills required for the role, ask about the project's objectives, deliverables, and key performance indicators (KPIs). This helps you understand the broader context and identify candidates who not only have the necessary skills but also align with the project's goals.

2. Clarify why the role exists.

Understanding the reason behind the role's creation is essential. Is it a backfill, a new initiative, or a growth opportunity? This information helps you tailor your search and identify candidates who fit the specific needs of the role.

3. Ask about past candidate mismatches.

Discussing past candidate mismatches can provide valuable insights into what didn't work previously. This helps you avoid similar mistakes and identify potential areas of concern, ensuring that you present candidates who are a better fit for the role.

4. Confirm deal breakers and preferred communication styles.

Deal breakers are non-negotiable factors that can make or break a candidate's suitability for the role. Confirming these requirements beforehand ensures that you don't waste time on candidates who don't meet the essential requirements. Additionally, understanding the preferred communication style helps you match your approach to the hiring manager's needs.

5. Discuss submission format preferences and an ideal number of resumes.

Understanding the hiring manager's preferences for submission formats and the ideal number of resumes helps you submit your application without confusion. This ensures that you provide the right amount of information in the most suitable format, making it easier for the hiring manager to review and make decisions.

Proactive Partnership Strategies

To build a strong partnership with the hiring manager, consider the following strategies:

1. Share real-time market feedback on candidate availability

Providing real-time market feedback on candidate availability helps set realistic expectations and demonstrates your expertise in the field. This transparency enables the hiring manager to make informed decisions and adjust their expectations accordingly.

2. Set realistic timelines based on domain and skill demand

Setting realistic timelines based on the domain and skill demand ensures that the hiring manager has a clear understanding of the recruitment process's duration. This helps manage expectations and avoids unnecessary delays.

3. Offer alternatives

Offering alternatives, such as "If we can't find X, would Y work temporarily?" demonstrates your problem-solving skills and willingness to adapt. This approach helps find creative solutions and ensures that the hiring manager's needs are met. It also reduces dependency on a particular thing or person and mitigates some unexpected misfortunes, as there are alternatives to fall back on in case one goes wrong.

4. Use intake notes and tech stack summaries

Using intake notes and tech stack summaries helps avoid repeat explanations and ensures that you have a thorough understanding of the role's requirements. This approach saves time and demonstrates your attention to detail.

Benefits of High-Impact Intake Calls

Conducting high-impact intake calls and implementing proactive partnership strategies can lead to the following:

1. Improved candidate quality
2. Increased efficiency in the recruitment process
3. Enhanced collaboration with hiring managers
4. Better alignment with project goals and objectives
5. Increased client satisfaction

Case Example: Rebuilding Trust through Structured Intake Calls

A recruiter faced challenges in maintaining the hiring manager's trust due to Mismatched submittals and a Lack of understanding of the role's requirements.

To address this, the Recruiter Initiated a structured intake call, Discussed project goals, role requirements, and potential red flags, and sent a summary of agreed-upon expectations, including ideal screening questions and red flags.

Outcome

The structured intake call and summary led to the following:

1. Drastically improved submittals
2. Rebuilt trust with the hiring manager

Key Takeaways

1. Structured intake calls can significantly improve understanding and alignment.
2. Clear communication and documentation of expectations are crucial.
3. A proactive approach can rebuild trust and drive successful outcomes.

This example underscores the importance of effective communication and structured processes in the recruitment process.

NOTE: This example is scenario-based training and learning, which helps us understand the scope better. The scope can be much deeper and more precise.

Self-Assessment

1. What are two pain points hiring managers commonly experience with recruiters?
2. Why is understanding project goals important in intake calls?
3. What should you do if a hiring manager mentions a candidate dropped out at the last minute?
4. How can you avoid asking repetitive questions during multiple intakes?
5. Name one way to manage hiring manager expectations proactively.

Reflective Prompt: Think of a past intake call. Did you ask more about the project or just the skills? What would you do differently today to make the intake more strategic and consultative?

CHAPTER FIVE

THE POWER OF COMMUNICATION AND SOFT SKILLS

Objectives of this Chapter:

- Understand how soft skills impact global recruitment success
- Learn effective communication strategies for working with U.S.-based clients and candidates
- Master email etiquette, tone, and followup practices
- Recognize cultural differences and how to navigate them with professionalism
- Build trust and credibility through empathy, clarity, and consistency

THE IMPORTANCE OF SOFT SKILLS IN RECRUITMENT.

In today's competitive job market, having the right technical skills is essential, but it is not enough. Soft skills, often referred to as people skills or interpersonal skills, play a crucial role in determining a candidate's success in a role. Here is why soft skills matter:

Hard Skills VS Soft Skills

Hard skills are technical skills that are specific to a job or industry. They are often easy to quantify and measure. Soft skills, on the other hand, are non-technical skills that relate to how you work with others, communicate, and approach challenges.

Why Soft Skills Matter

Soft skills are essential because they:

1. Enhance teamwork and collaboration: Soft skills, such as communication, empathy, and conflict resolution, help team members work together effectively.
2. Improve communication: Soft skills, such as active listening, clarity, and adaptability, ensure that messages are conveyed effectively.

3. Foster problem-solving problem-solving and adaptability: Soft skills, such as creativity, critical thinking, and resilience, help employees navigate complex challenges.
4. Drive customer satisfaction: Soft skills, such as empathy and patience, combined with customer service skills, ensure that customers receive excellent service.
5. Support leadership and management: Soft skills, such as leadership, coaching, and mentoring, help managers motivate and develop their teams.

Benefits of Soft Skills in Recruitment

By prioritizing soft skills in recruitment, organizations can:

1. Reduce turnover rates: Employees with strong soft skills are more likely to align with the company culture and work effectively with others.
2. Enhance team dynamics: Soft skills such as teamwork, communication, and empathy foster a positive and productive work environment.
3. Enhance customer satisfaction: Employees with strong soft skills deliver better customer service, resulting in increased customer satisfaction and loyalty.
4. Increase productivity: Soft skills, such as time management, prioritization, and adaptability, enable employees to manage their workload and meet deadlines effectively.

How to assess Soft Skills in Recruitment

To assess soft skills in recruitment, consider:

1. Behavioral interviews: Ask candidates to provide specific examples of how they have demonstrated soft skills in the past.
2. Role playingRole-playing exercises: Use role-playing role-playing exercises to assess a candidate's communication, problem-solving, and teamwork skills. Ask them a few questions like "If this happens to a job you are working on, and it needs to be delivered in the next two hours, how can you tell your client about what happened without making him angry, and how do you intend to convince the stakeholders that you are still capable of fixing your mistakes?"

3. Reference checks: Ask references about the Candidate's soft skills and how they have demonstrated them in previous roles.

Soft skills are essential for success in today's workplace. By prioritizing soft skills in recruitment, organizations can build strong, productive teams that consistently deliver exceptional results. Remember, hard skills might get your foot in the door, but soft skills keep it open. A very qualified candidate can secure a job with her qualifications, but when she is rude to her boss or clients, hates teamwork with her colleagues, and does not possess the right attitude to work, there is a high chance that she will lose the job.

Ultimately, it is not just about finding candidates with the right technical skills; it is about identifying individuals who can collaborate effectively, communicate clearly, and tackle challenges with creativity and resilience. By focusing on soft skills, organizations can establish a solid foundation for success.

THE IMPACT OF SOFT SKILLS ON GLOBAL RECRUITMENT SUCCESS.

In today's interconnected and globalized business landscape, soft skills have become a crucial factor in determining the success of recruitment efforts. As companies expand their operations across borders, they are finally realizing that technical skills alone are insufficient to guarantee success. Soft skills, such as communication, teamwork, and adaptability, are essential for navigating the complexities of global business.

Why Soft Skills Matter in Global Recruitment.

Earlier in this chapter, we defined soft skills. But for the sake of this context, we would be defining it again. Soft skills are non-technical skills that relate to how you work with others, communicate, and approach challenges.

Soft skills are vital in global recruitment because they:

1. Facilitate cross-cultural communication: Soft skills like empathy, active listening, and clarity help bridge cultural gaps and ensure effective communication across diverse teams.
2. Enable global teamwork: Soft skills such as collaboration, flexibility, and conflict resolution enable teams to work together seamlessly despite geographical and cultural differences.

3. Support adaptability in diverse environments: Soft skills such as resilience, open-mindedness, and adaptability enable employees to navigate unfamiliar environments and adjust to new cultural norms.
4. Foster innovation and creativity: Soft skills like creativity, critical thinking, and problem-solving enable employees to approach challenges from unique perspectives, driving innovation and growth.

Benefits of Soft Skills in Global Recruitment

By prioritizing soft skills in global recruitment, organizations can:

1. Enhance collaboration and teamwork: Soft skills enable global teams to work together effectively, leading to improved outcomes and increased productivity.
2. Enhance customer satisfaction: Soft skills, such as empathy, customer service, and effective communication, ensure that customers receive excellent service, regardless of their location or cultural background.
3. Increase employee engagement and retention: Soft skills, such as leadership, coaching, and mentoring, enable managers to motivate and develop their teams, thereby reducing turnover rates and enhancing employee satisfaction.
4. Drive business growth and innovation: Soft skills, such as creativity, problem-solving, and adaptability, enable employees to navigate complex global markets and identify new opportunities for growth.

Challenges of Assessing Soft Skills in Global Recruitment.

For something that draws a lot of people together, there would be some challenges. Assessing soft skills in global recruitment can be challenging due to:

1. Cultural differences: Soft skills can be perceived differently across cultures, making it essential to consider cultural nuances when assessing candidates.
2. Language barriers: Language differences can create barriers to effective communication, making it crucial to assess candidates' language skills and ability to communicate across linguistic and cultural boundaries.
3. Diverse work environments: Soft skills can be demonstrated differently in various work environments, requiring recruiters to consider the specific context and cultural norms of the role.

Best Practices for Assessing Soft Skills in Global Recruitment

To assess soft skills effectively in global recruitment:

1. Use culturally sensitive assessment tools: Develop assessment tools that consider cultural differences and nuances.
2. Conduct behavioural interviews: Use behavioural interviews to assess candidates' past experiences and behaviours in diverse cultural contexts.
3. Utilize scenario-based assessments: Use scenario-based assessments to evaluate candidates' problem-solving and decision-making skills in global business contexts.
4. Involve diverse interview panels: Involve interview panels from diverse cultural backgrounds to provide a more comprehensive assessment of candidates' soft skills.

Most of these have been explained earlier, so it is advisable to review them again to prevent repetition.

EFFECTIVE COMMUNICATION STRATEGIES FOR WORKING WITH U.S.-BASED CLIENTS AND CANDIDATES.

Communication is the process of exchanging information, ideas, thoughts, and messages between individuals, groups, or organizations through various channels, including verbal (spoken words), nonverbal (body language and facial expressions), written (emails, letters, and texts), and visual (images and videos). It is the act of sharing or exchanging information, ideas, or messages between people through talking, writing, or other forms of expression. It is how we connect, understand, and convey thoughts and feelings to each other.

Communication is considered adequate when it involves sending clear and concise messages, engaging in active listening and understanding, and providing feedback and clarification.

When working with U.S.-based clients and candidates, effective communication is essential for building strong relationships, preventing misunderstandings, and achieving successful outcomes. Here are some strategies to help you communicate effectively:

1. Understand Cultural Nuances

The United States is a culturally diverse country, and understanding these nuances can help you organize or reform the way you communicate with them. For example:

- Directness and assertiveness are often valued in U.S. business culture.

Building rapport and establishing trust are crucial for maintaining successful relationships.

2. Use Clear and Concise Language

Clear and concise language helps avoid misunderstandings and ensures that your message is conveyed effectively:

- Avoid using jargon or technical terms that may be unfamiliar to your audience as you do not want to confuse them.
- Use simple, straightforward language to convey complex ideas. Do not beat around the bush.

3. Be Responsive and Reliable

U.S.-based clients and candidates often expect prompt responses and reliable followup:

- Respond to emails, calls, and messages promptly.
- Follow up on commitments and deadlines to build trust and credibility.

4. Use Technology Effectively

Technology can facilitate communication, but it is essential to use it effectively:

- Use video conferencing tools for face-to-face communication.
- Leverage collaboration tools, such as Slack or Trello, to facilitate team communication.

5. Be Adaptable and Flexible

U.S.-based clients and candidates may have different communication styles and preferences, so you have to:

- Be open to adjusting your communication approach to meet their needs.

- Be flexible with scheduling and communication channels.

6. Show Appreciation and Gratitude

This is an essential strategy. Expressing appreciation and gratitude can go a long way in building strong relationships:

- Show appreciation for clients' and candidates' time and consideration.
- Express gratitude for their business or partnership no matter how little you might think it is to them.

Best Practices for Communication

Some additional best practices to keep in mind:

- Use active listening skills: Pay attention to what the other person is saying and respond thoughtfully.
- Ask clarifying questions: Ensure you understand the other person's needs and concerns. Do not assume things. Ensure you understand everything from the client's perspective and try to establish a rapport with them to prevent future misunderstandings.
- Be mindful of time zones: Schedule calls and meetings at convenient times for U.S.-based clients and candidates.
- Use professional language: Avoid using slang or overly casual language in professional communication.

Effective communication is key to building strong relationships, especially with U.S.-basedU.S.-based clients and candidates. By understanding cultural nuances, using clear and concise language, being responsive and reliable, and leveraging technology effectively, you can drive successful outcomes and establish yourself as a trusted partner.

THE ART OF COMMUNICATION IN GLOBAL RECRUITMENT.

In global recruitment, communication serves as the bridge that connects talented individuals from diverse backgrounds to opportunities that align with their skills and aspirations. However, effective communication goes beyond just language proficiency. It is about about understanding the complexities of human interaction, cultural differences, and individual expectations.

Clarity: The Foundation of Effective Communication

Clarity is essential in global recruitment communication. It involves:

1. Clear job descriptions: Ensure that job requirements and expectations are well-defined and easily understood.
2. Transparent communication: Provide regular updates on the recruitment process and be open about the company's culture, values, and goals.
3. Simple language: Avoid using jargon or technical terms that may be unfamiliar to non-native speakers.

Empathy: Building Trust and Rapport

Empathy is critical in building trust and rapport with candidates and clients from diverse cultural backgrounds. It involves:

1. Active listening: Pay attention to the Candidate's needs, concerns, and aspirations.
2. Cultural understanding: Be sensitive to cultural differences and other things that may impact communication.
3. Personalized approach: Tailor your communication approach to meet the individual's specific needs and preferences.

Timing: The Importance of Responsiveness

Timing is crucial in global recruitment communication. It involves:

1. Prompt responses: Respond to emails, calls, and messages promptly to show that you value the Candidate's time.
2. Scheduled updates: Provide regular updates on the recruitment process to keep candidates informed.
3. Flexibility: Be flexible with scheduling and communication channels to accommodate different time zones and preferences.

Tone: Setting the Right Impression

Tone is essential in global recruitment communication. It is the way you approach people and the kind of body language and words you use on them. It involves:

1. Professional tone: Maintain a professional tone in all communication, whether it is an email, phone call, or video conference.

2. Friendly and approachable: Be friendly and approachable in your communication while still maintaining professionalism.
3. Cultural sensitivity: Be mindful of cultural differences that may impact tone and communication style.

Adaptability: The Key to Success

Adaptability is critical in global recruitment communication. It is the ability to adapt quickly to new things. It involves:

1. Flexibility in communication style: Adapt your communication style to the individual's needs and preferences.
2. Cultural awareness: Be aware of cultural differences and nuances that may impact communication.
3. Open-mindedness: Be open-minded and willing to learn from different cultures and perspectives.

Connecting Humans Across Cultures and Expectations

In global recruitment, you are not just connecting resumes to jobs; you are connecting humans across cultures and expectations. It is about:

1. Understanding individual needs: Understand the individual needs, aspirations, and expectations of candidates and clients.
2. Building relationships: Build strong relationships with candidates and clients based on trust, empathy, and understanding.
3. Facilitating cultural exchange: Facilitate cultural exchange and understanding between candidates, clients, and colleagues from diverse backgrounds.

By mastering the art of communication in global recruitment, you can build strong relationships, drive successful outcomes, and make a positive impact on the lives of candidates and clients from around the world.

Effective Communication Channels in Recruitment

In today's fast-paced recruitment landscape, effective communication is crucial for building strong relationships with candidates, clients, and colleagues. Various communication channels play a vital role in facilitating seamless interaction and driving successful outcomes. Let's dive into the key communication channels and explore their significance:

1. Email: The Foundation of Professional Communication

Email remains the most commonly used communication channel in recruitment. It is essential to ensure that emails are:

- Clear: Easy to understand, with a clear purpose and call to action.
- Concise: Brief and to the point, avoiding unnecessary details.
- Grammatically clean: Well written, with proper grammar, spelling, and punctuation.

Email is ideal for:

- Sending job descriptions, interview schedules, and offer letters.
- Sharing updates on the recruitment process.
- Following up with candidates and clients.

2. Phone: Building Relationships and Closing Deals

Phone calls are an essential communication channel in recruitment, particularly for:

- Initial screening: Phone calls help assess a candidate's communication skills, tone, and enthusiasm.
- Offer negotiations: Phone calls facilitate discussions about salary, benefits, and other employment details.
- Closing deals: Phone calls can seal the deal and finalize agreements.

3. Messaging Apps: Day-to-Day Coordination

Messaging apps like WhatsApp, Skype, and MS Teams are perfect for:

- Quick updates: Sharing brief updates, reminders, or notifications.
- Day-to-day coordination: Discussing logistics, scheduling interviews, or sharing documents.
- Informal communication: Building rapport and establishing a personal connection with candidates and colleagues.

4. Video Calls: Building Relationships and Conducting Assessments

Video calls are crucial for:

- Building relationships: Face-to-face interactions help establish trust and rapport with candidates and clients.

- Conducting assessments: Video calls enable recruiters to assess a candidate's communication skills, body language, and personality.
- Remote interviews: Video calls facilitate remote interviews, reducing the need for in-person meetings. In cases where the distance between the parties is great, they can conduct video calls efficiently using platforms like Skype, Zoom, Google Meet, and even WhatsApp.

Other Communication Channels

In addition to the above channels, other communication channels that can be used in recruitment include:

- Social media: Utilize platforms like LinkedIn, Twitter, or Facebook to engage with candidates, share job openings, and promote the company's brand.
- Project management tools: Tools like Trello, Asana, or Basecamp help manage workflows, assign tasks, and track progress.
- Instant messaging platforms: Platforms like Slack or Google Workspace enable real-time communication and collaboration among team members.
- Video conferencing platforms: Platforms like Zoom or Google Meet facilitate virtual meetings, interviews, and presentations.

To maximize the effectiveness of communication channels:

- Choose the right channel: Select the most suitable channel for the message, audience, and purpose. For example, if you want to set a reminder, you could do it via WhatsApp instead of having to start a conference call.
- Be clear and concise: Ensure that messages are easy to understand and free of ambiguity.
- Be responsive: Respond promptly to messages, emails, and calls.
- Use technology strategically: Leverage technology to streamline communication, automate tasks, and enhance productivity.

By mastering the art of communication and utilising the proper channels, recruiters can cultivate strong relationships, drive successful outcomes, and leave a lasting impression on both candidates and clients.

EMAIL ETIQUETTE ESSENTIALS FOR RECRUITMENT PROFESSIONALS.

Email etiquette is crucial in the recruitment industry, where professionalism, clarity, and attention to detail are paramount. A well-crafted email can make a lasting impression on candidates, clients, and colleagues, while a poorly written one can lead to misunderstandings, miscommunications, and lost opportunities.

Email etiquette includes:

1. Use Professional Subject Lines

A professional subject line is essential for several reasons:

- Clarity: It indicates the purpose of the email, helping the recipient prioritize their inbox effectively.
- Relevance: It ensures that the email is relevant to the recipient, increasing the likelihood of a response.
- Organization: It helps the recipient quickly identify the email's context and categorize it accordingly.

The best ways to write your subject lines include:

- Be specific: Use specific keywords related to the job, Candidate, or client.
- Keep it concise: Aim for a subject line that is no more than 5-7 words.
- Use relevant details: Include relevant details such as the Candidate's name, job title, or client name.

Example: "Java Developer Submission – John D. – Healthcare Client"

2. Keep Intros Short and Focused

A brief and focused introduction is vital for grabbing the reader's attention and conveying the email's purpose:

- Be concise: Keep the introduction to 1-2 sentences, focusing on the main point.
- Be clear: Clearly state the purpose of the email, avoiding ambiguity.
- Use a professional tone: Use a professional tone and language, avoiding jargon and slang.

Example: "Dear [Client], I am pleased to submit the resume of John D., a highly qualified Java Developer, for your consideration."

3. Use Bullet Points for Candidate Summaries

Bullet points are an effective way to present candidate summaries, making it easy for the reader to scan and understand the information:

- Highlight key skills: Use bullet points to highlight the Candidate's key skills, experience, and achievements.
- Keep it concise: Keep each bullet point brief, focusing on the most critical information.
- Use action verbs: Use action verbs like "Managed," "Developed," and "Improved" to describe the Candidate's experience.

Example:

- 5+ years of experience in Java development
- Proven track record of delivering high-quality software solutions
- Strong expertise in Agile methodologies and Scrum framework

4. Always Proofread for Grammar and Tone

Proofreading is essential for ensuring that the email is free of errors and conveys the intended tone:

- Grammar and spelling: Check for grammar, spelling, and punctuation errors.
- Tone and language: Ensure that the tone and language are professional and respectful.
- Clarity and concision: Verify that the email is clear, concise, and easy to understand.

Best practices for proofreading include:

- Take a break: Take a break before proofreading to approach the email with fresh eyes and mind, not a tired one.
- Read aloud: Read the email aloud to detect any awkward phrasing or tone.
- Use tools: Use grammar and spell check tools to identify errors. Example: Grammarly app.

By following these email etiquette essentials, recruitment professionals can ensure that their emails are professional, effective, and well-received by candidates, clients, and colleagues.

VOICE AND TONE GUIDELINES FOR EFFECTIVE COMMUNICATION.

In the fast-paced recruitment world, effective communication is crucial for building strong relationships with candidates, clients, and colleagues. Voice and tone play a significant role in conveying professionalism, empathy, and enthusiasm. So now, we shall dive into the voice and tone guidelines that can help you communicate effectively:

1. Speak Clearly and Avoid Filler Words

Speaking clearly and avoiding filler words is essential for conveying confidence and professionalism:

- Enunciate: Pronounce words clearly and correctly, avoiding mumbling or rushing.
- Avoid filler words: Refrain from using filler words like "um," "ah," or "you know" that can detract from your message.
- Pause: Pause briefly to collect your thoughts, allowing for a more composed and thoughtful response.

2. Adjust Speed for Accents (Yours and Theirs)

Adjusting your speaking speed can help ensure that your message is conveyed clearly, especially when you are communicating with individuals with different accents:

- Be mindful of your accent: Be aware of your accent and adjust your speaking speed accordingly.
- Listen actively: Pay attention to the speaker's accent and adjust your listening speed to ensure understanding.
- Clarify: If necessary, ask for clarification or repeat information to ensure mutual understanding.

3. Practice Empathy When Discussing Rate, Rejection, or Interview Feedback

Empathy is crucial when discussing sensitive topics like rate, rejection, or interview feedback and some of the ways you can show empathy are:

- Be compassionate: Show understanding and compassion when discussing challenging topics.
- Listen actively: Pay attention to the speaker's concerns and respond thoughtfully.
- Provide constructive feedback: Offer specific, actionable feedback that can help the Candidate grow.

4. Use Scripts Initially, But Adapt Naturally

Using scripts can help you prepare for everyday conversations, but it is essential to adapt naturally to each situation:

- Prepare: Develop scripts for everyday conversations, such as interview questions or rejection notifications.
- Adapt: Adapt your script to fit the specific situation and Candidate, showing genuine interest and empathy.
- Be authentic: Be yourself, and don't be afraid to deviate from the script if the Conversation requires it.

Benefits of Effective Voice and Tone

Compelling voice and tone can have numerous benefits, including:

- Building trust: Clear and empathetic communication can help build trust with candidates and clients.
- Improving relationships: A professional and friendly tone can foster strong relationships and improve communication.
- Enhancing reputation: Consistent and effective communication can enhance your reputation as a recruitment professional.

To master compelling voice and tone:

- Practice: Record yourself and practice your communication skills.
- Get feedback: Seek feedback from colleagues, candidates, or mentors.
- Be self-aware: Recognize your strengths and areas for improvement.

By following these voice and tone guidelines, you can communicate effectively, build strong relationships, and drive successful outcomes in the recruitment industry.

FOLLOWUP FRAMEWORK FOR EFFECTIVE COMMUNICATION.

Earlier, we discussed the benefits of effective communication with clients, candidates, and others. After effective communication, the next step is an adequate followup, which is what we will be discussing here. Adequate followup is crucial for building strong relationships with candidates and clients. A well-structured follow-up framework can help you stay on top of your mind, manage expectations, and drive successful outcomes. Let's dive into the key components of a followup framework:

1. Acknowledge Receipt or Next Steps

Acknowledging receipt or next steps is essential for setting the tone for your communication:

- Respond promptly: Respond to emails, calls, or messages promptly.
- Confirm receipt: Confirm receipt of resumes, applications, or other vital documents.
- Outline next steps: Clearly outline the next steps in the process, including timelines and expectations, to ensure a smooth transition.

Example: "Thank you for submitting your resume for the [Job Title] position. We will review your application and get back to you within [Timeframe] with an update."

2. Set Time-Based Expectations

Setting time-based expectations helps manage candidate and client expectations:

- Provide specific timelines: Provide specific timelines for each stage of the process.
- Be realistic: Be realistic about the time it takes to complete each stage.
- Communicate proactively: If there are any delays or changes, communicate them promptly and effectively.

Example: "We anticipate conducting interviews for the [Job Title] position within the next [Timeframe]. You will receive an email with the interview schedule and details."

3. Remind Without Pestering

Reminding without pestering is a delicate balance:

- Schedule reminders: Schedule reminders for followup calls or emails.
- Be considerate: Be considerate of the candidate's or client's time and preferences.
- Add value: Add value to each followup interaction, whether it's providing an update or sharing relevant information.

Example: "Hi [Candidate], I just wanted to follow up on the status of your application. We are still reviewing resumes and will be in touch soon."

4. Thank Clients/Candidates Often

Expressing gratitude can go a long way in building strong relationships:

- Show appreciation: Show appreciation for the Candidate's interest in the role or the client's business.
- Be sincere: Be sincere in your gratitude, avoiding generic or insincere expressions.
- Celebrate milestones: Celebrate milestones, such as a successful placement or a new hire's start date.

Example: "Thank you for considering our recruitment services. We are thrilled to have placed [Candidate] in the [Job Title] role. Please don't hesitate to reach out if you need any further assistance."

Benefits of a Followup Framework

A followup framework can have numerous benefits, including:

- Improved communication: Clear and timely communication can improve relationships and drive successful outcomes.
- Increased trust: Consistent followup can build trust and credibility with candidates and clients.
- Enhanced reputation: A well-structured follow-up framework can enhance your reputation as a recruitment professional.

To master adequate followup:

- Be proactive: Be proactive in your communication, anticipating needs and concerns.
- Use technology: Use technology to streamline followup, such as automated email reminders or CRM systems.
- Personalise: Tailor your follow-up interactions to each Candidate or client.

By implementing a followup framework, you can build strong relationships, drive successful outcomes, and establish yourself as a trusted and effective recruitment professional.

GLOBAL COMMUNICATION STYLE COMPARISON.

	DIRECTNESS	FEEDBACK STYLE	TIME EXPECTATIONS	TONE
North America	Direct and straightforward	Constructive, specific and solution-oriented.	Punctuality and efficiency are valued	Friendly, informal and approachable.
Asia	More indirect, polite and respectful.	Feedback is often implicit, avoiding direct criticism.	Relationships and hierarchy are essential and time-flexible.	Formal, respectful and courteous
Europe	It varies by country but is generally direct.	Feedback is often critical but constructive.	Punctuality is valued, but flexibility is acceptable.	Varies by country, but often formal and professional
Africa	Varies by country and culture, but often indirect	Feedback is often given in a communal or group setting.	Time flexible, with emphasis on relationships	Warm, respectful and community-oriented

CULTURAL SENSITIVITY TIPS FOR GLOBAL RECRUITMENT

In today's recruitment landscape, cultural sensitivity is essential for building strong relationships with candidates and clients from diverse backgrounds. Being aware of cultural differences and adapting your approach can help you navigate complex situations and avoid unintended misunderstandings. Let us dive into some essential cultural sensitivity tips:

1. Be Mindful of Public Holidays and Time Zones

Being mindful of public holidays and time zones is essential for respecting candidates' and clients' cultural and geographical differences:

- Research public holidays: Research public holidays and observances in different countries and cultures.
- Schedule accordingly: Schedule meetings, calls, and deadlines around these holidays and time zones to ensure optimal coordination and communication.
- Be flexible: Be flexible with scheduling to accommodate different time zones and work styles.

2. Avoid Idioms or Slang

Avoiding idioms or slang can help prevent misunderstandings and ensure clear communication:

- Use precise language: Use clear and concise language that is easy to understand.
- Avoid cultural references: Avoid cultural references or idioms that may not be familiar to non-native speakers.
- Define technical terms: Define technical terms or industry-specific jargon.

3. Respect Differences in Assertiveness and Formality

Respecting differences in assertiveness and formality can help you build stronger relationships with candidates and clients:

- Understand cultural norms: Understand cultural norms around assertiveness and formality.
- Adapt your approach: Adapt your approach to fit the cultural context and individual preferences.
- Be sensitive to nonverbal cues: Be sensitive to nonverbal cues, such as body language and tone of voice.

4. Clarify Expectations Without Assuming

Clarifying expectations without assuming can help prevent misunderstandings and ensure successful outcomes:

- Ask questions: Clarify expectations and ensure mutual understanding by asking questions.
- Avoid assumptions: Avoid making assumptions about cultural norms or individual preferences.
- Confirm understanding: Confirm understanding and agreement on expectations and next steps.

Benefits of Cultural Sensitivity

Cultural sensitivity can have numerous benefits, including:

- Improved relationships: Culturally sensitive approaches can help build stronger relationships with candidates and clients.
- Increased trust: Being respectful of cultural differences can increase confidence and credibility.
- Enhanced reputation: Demonstrating cultural sensitivity can enhance your reputation as a recruitment professional.

Best Practices for Cultural Sensitivity

To master cultural sensitivity:

- Educate yourself: Educate yourself about different cultures and customs.
- Seek feedback: Seek feedback from candidates, clients, and colleagues from diverse backgrounds.
- Be open-minded: Be open-minded and willing to adapt your approach to fit different cultural contexts.

By following these cultural sensitivity tips, you can build stronger relationships, drive successful outcomes, and establish yourself as a respected and effective recruitment professional in the global market.

Case Example: A recruiter from Oceania realized that their formal emails were coming across as cold to their client in the U.S. Northeast. They decided to adjust their tone by using a friendly greeting and adding a short personal note, such as "Hope your week is going well!" This change had an impact on the client's response. By adding a bit of warmth and cheerfulness to the email, the recruiter was able to build a stronger rapport with the client.

The client's responses became more candid and forthcoming, resulting in improved overall communication. The recruiter discovered that a subtle adjustment in tone and language can significantly impact how messages are perceived. By being more mindful of the client's cultural background and communication style, the recruiter was able to understand and adapt to better meet the client's needs.

This experience highlights the importance of being adaptable and sensitive to cultural differences in communication. By taking the time to understand the client's preferences and adjusting their approach accordingly, the recruiter was able to build a stronger relationship and achieve better outcomes. The recruiter's willingness to adapt and be more personal in their communication helped to establish trust and credibility with the client, ultimately leading to more effective collaboration and successful placements.

NOTE: This example is scenario-based for training and learning purposes to help us understand the scope better. The real scope can be much deeper and more precise.

Self-Assessment #5

1. Why are soft skills as important as technical skills in global recruiting?
2. What's one way to improve clarity in email communication?
3. How can voice tone impact your client or Candidate call?
4. Name two platforms where soft communication skills are key.
5. Why is it important to be culturally sensitive when working with U.S. clients?

Reflective Prompt: Recall a time when your communication style helped or hurt your professional interaction. What would you keep, and what would you change next time?

CHAPTER SIX

COMPENSATION MODELS AND RATE STRUCTURES IN THE US

Objectives of this Chapter:

- Understand the different compensation types in U.S. recruitment (W2, C2C, 1099)
- Learn the difference between salary and hourly models
- Gain awareness of how location and cost of living affect compensation
- Know how to discuss bill rates, pay rates, and margins with clarity
- Build the confidence to negotiate compensation in alignment with client and candidate expectations

DIFFERENT COMPENSATION TYPES IN THE U.S

Compensation refers to the rewards or payments an organization provides to its employees in exchange for their work. This can include monetary compensation, such as salary, wages, commissions, and bonuses, as well as non-monetary benefits, including health insurance, retirement plans, paid time off, and flexible working hours.

In this context of recruitment, compensation is something used by management to achieve various purposes, including:

- Attracting and retaining qualified employees: Offering competitive compensation packages to draw in top talent and keep them engaged because, of course, everyone wants to work in a place where they are adequately paid. For example, if an influencer sees two job openings with the same workload but different payment amounts, she would go for the higher-paying job.
- Boosting morale and job satisfaction: Ensuring fair compensation to maintain employee satisfaction and motivation. Whenever they feel stressed, they can recall how much they are being paid and how no other job can offer them the same compensation, which motivates them to continue working and makes them very happy and satisfied.

- Rewarding performance: Using compensation to recognize and incentivize exceptional work. Everyone loves to be seen or noticed for their efforts.

Compensation can be categorized into :

- Direct Compensation: Financial rewards like base pay, overtime pay, commissions, and bonuses.
- Indirect Compensation: Non-monetary benefits like health insurance, stock options, and flexible working hours.
- Total Compensation: The combination of direct and indirect compensation, forming a comprehensive package.

Effective compensation strategies are essential for organizations to attract, retain, and motivate employees, ultimately driving business success.

Now, Let's break down the different compensation types in U.S. recruitment, specifically W2, C2C (Corp-to-Corp), and 1099. Understanding these compensation structures is crucial for both employers and employees to navigate the complex recruitment landscape.

W2 Employees

W2 employees are traditional employees who receive a salary or wages from an employer. The employer withholds taxes, including Social Security and Medicare taxes, from the employee's paycheck. W2 employees typically receive benefits, such as health insurance, retirement plans, and paid time off.

- Pros for Employers:
- Control over work schedules and assignments
- Eligibility for benefits, reducing administrative burden
- Can be more cost-effective for long-term engagements
- Cons for Employers:
- Payroll taxes and benefits administration
- Potential liability for employee actions
- Limited flexibility in payment structures

1099 Contractors

One thousand ninety-nine contractors, also known as independent contractors, are self-employed individuals who work on a project-by-project basis. They are responsible for their taxes, including self-employment taxes, and typically do not receive benefits from the employer.

- Pros for Employers:
- Flexibility in hiring and project management
- Reduced liability for taxes and benefits
- Can be more cost-effective for short-term or specialized projects
- Cons for Employers:
- Less control over work schedules and assignments
- Potential misclassification risks
- Limited ability to provide benefits or training

C2C (Corp-to-Corp) Contractors

C2C contractors involve a contractual agreement between two companies, where one company provides services to the other through an individual or team. The contractor company invoices the client company, and the contractor is responsible for their taxes and benefits.

- Pros for Employers:
- Reduced liability for taxes and benefits
- Flexibility in project management and engagement
- Can be beneficial for specialized or high-demand skills
- Cons for Employers:
- Potential complexity in contract management
- Limited control over work schedules and assignments
- May require more administrative effort

When deciding between W2, 1099, and C2C compensation structures, employers should consider factors such as:

- Project duration and scope: Short-term projects benefit from 1099 or C2C arrangements, while long-term projects are more suitable for W2 employees.

- Control and flexibility: W2 employees offer more control, while 1099 and C2C contractors provide more flexibility.
- Taxes and benefits: Employers must consider the administrative burden and costs associated with each compensation structure.

Ultimately, understanding the differences between W2, 1099, and C2C compensation structures can help employers make informed decisions about their recruitment strategies and workforce management.

The Differences between a Salary model and an hourly model.

The primary difference between salary and hourly models lies in how employees are compensated for their work.

Salary Model

In a salary model, employees are paid a fixed annual amount, usually divided into regular pay periods (e.g., bi-weekly or monthly). Salaried employees are often exempt from overtime pay and are expected to work a standard number of hours per week.

Advantages of this model:

It provides a Predictable income for employees, meaning they know the amount of money they can expect at the end of the month, allowing them to make plans accordingly.

- It encourages employees to focus on results rather than hours worked
- Can be beneficial for professional or managerial roles
- Disadvantages:
- It could lead to burnout if employees work excessive hours without additional compensation
- Can be inflexible for employees with varying work schedules.

Hourly Model

In an hourly model, employees are paid for each hour they work. Hourly employees are often entitled to overtime pay for hours worked beyond a standard threshold (e.g., 40 hours per week).

- Advantages of the Hourly model:

- It provides flexibility for employees with varying work schedules
- Can be beneficial for roles with fluctuating workloads or variable hours
- Overtime pay can incentivize employees to work extra hours
- -Disadvantages:
- Income may be unpredictable for employees
- Can create administrative burdens for tracking hours worked

The choice between salary and hourly models depends on the specific needs and goals of the organization, as well as the nature of the work being performed. Some roles may be better suited to one model over the other, and some employees may prefer the predictability of a salary or the flexibility of hourly pay.

The salary model is often used for professional, managerial, and administrative roles where employees are expected to work a standard number of hours per week. Some common job types that are often salaried include:

- Professional roles:
- Software engineers
- Lawyers
- Doctors
- Accountants
- Managerial roles:
- Department managers
- Team leads
- Senior executives
- Administrative roles:
- Office managers
- Human resources managers
- Financial analysts

The hourly model, on the other hand, is often used for roles where the number of hours worked can vary significantly or where the work is more project-based. Some common job types that are often hourly include:

Service industry roles:

- Retail sales associates
- Restaurant staff

- Hotel staff
- Skilled trades:
- Electricians
- Plumbers
- Carpenters
- Gig economy roles:
- Rideshare drivers
- Food delivery drivers
- Freelance writers or designers
- Part-time or temporary roles:
- Part-time retail staff
- Temporary office workers
- Seasonal workers

Keep in mind that these are general examples, and the specific job types that are salaried or hourly can vary depending on the industry, company, and location.

THE IMPACT OF LOCATION AND COST OF LIVING ON COMPENSATION.

The impact of location and cost of living on compensation is an essential consideration for both employers and employees. Let's dive into the details.

Location Based Compensation

Location plays a significant role in determining compensation packages. The cost of living varies significantly across different regions, cities, and even neighbourhoods. Employers often adjust salaries and benefits to reflect the local cost of living, ensuring that employees can maintain a reasonable standard of living.

- Urban vs. rural areas: Cities tend to have a higher cost of living compared to rural areas. Employers may offer higher salaries in urban areas to compensate for the increased cost of housing, transportation, and other living expenses.
- Regional differences: Different regions have unique cost of living profiles. For example, the cost of living in New York City is significantly higher than in other parts of the country. Employers may adjust salaries accordingly to ensure that employees can afford the local cost of living.

- Industry-specific considerations: Certain industries, such as tech, may have different compensation standards in other locations. For example, a software engineer in San Francisco may command a higher salary than one in a smaller city.

Cost of Living Adjustments

Cost-of-living adjustments (COLAs) are changes made to salaries or benefits to reflect changes in the cost of living. COLAs can be applied to various aspects of compensation, including:

- Salary adjustments: Employers may adjust salaries to reflect changes in the cost of living, ensuring that employees' purchasing power remains relatively constant.
- Benefits adjustments: Employers may also adjust benefits, such as health insurance or retirement plans, to reflect changes in the cost of living.
- Relocation packages: When relocating employees to a new location, employers may offer relocation packages that include cost-of-living adjustments to help employees transition to the new location.

Impact on Employee Satisfaction and Retention

Location-based compensation and cost-of-living adjustments can have a significant impact on employee satisfaction and retention. When employees feel that their compensation package reflects the local cost of living, they are more likely to:

- Feel valued and appreciated: Employees who receive fair compensation packages are more likely to feel valued and appreciated by their employers.
- Be satisfied with their job: Employees who are happy with their compensation packages are more likely to be satisfied with their jobs overall.
- Stay with the company: Employees who feel that their compensation package is fair and competitive are more likely to stay with the company long-term.

What the Employers should do.

To ensure that compensation packages are competitive and reflective of the local cost of living, employers can:

- Conduct regular cost-of-living analyses: Employers can conduct regular analyses to determine the cost of living in their location and adjust compensation packages accordingly.
- Offer flexible benefits: Employers can provide flexible benefits that enable employees to select the options that best suit their needs and lifestyle.
- Provide transparent compensation information: Employers can provide transparent information about compensation packages and how they are determined, helping employees understand the value of their compensation.

By understanding the impact of location and cost of living on compensation, employers can create competitive and fair compensation packages that attract and retain top talent.

COST OF LIVING CONSIDERATIONS

The cost of living is a critical consideration when discussing salaries and compensation packages. It is essential to recognize that the same salary can have vastly different purchasing powers in other cities.

Regional Differences in Cost of Living

The cost of living varies significantly across different regions, cities, and even neighbourhoods. For example, $130,000 in San Francisco might not go as far as $130,000 in Dallas due to differences in housing costs, taxes, transportation, and other living expenses.

- Housing costs: The cost of housing is a significant factor in the overall cost of living. Cities like San Francisco, New York, and Los Angeles have notoriously high housing costs, while cities like Dallas, Denver, and Austin have relatively lower costs.
- Taxes: State and local taxes can also impact the cost of living. Some states have higher tax rates, while others have lower tax rates. This can affect the take-home pay and overall compensation package.
- Transportation costs: Transportation costs, including gas, parking, and public transportation, can vary significantly depending on the city.

Client Considerations

Clients are increasingly aware of the regional differences in cost of living and factor this into their budgeting decisions. They recognize that the same salary can have different implications in different cities.

- Budgeting: Clients consider the cost of living in a particular city when determining their budget for salaries and benefits.
- Talent attraction and retention: Clients also recognize the need to offer competitive salaries and benefits to attract and retain top talent in various cities.

Recruiter Considerations

Recruiters play a critical role in ensuring that rate discussions and expectations are aligned with the client's budget and the Candidate's needs. Therefore, recruiters need to factor in the cost of living when discussing salaries and compensation packages.

- Rate discussions: Recruiters should consider the cost of living when discussing rates with clients and candidates. They should ensure that the salary or hourly rate is competitive and reflects the local market conditions.
- Candidate expectations: Recruiters should also understand the Candidate's expectations and needs, including their salary requirements and lifestyle preferences.
- Market knowledge: Recruiters should have an in-depth understanding of the local market, including the cost of living, industry standards, and market rates.

To ensure that cost of living considerations are factored into rate discussions and expectations, recruiters can follow these practices:

- Research local market conditions: Recruiters should research local market conditions, including the cost of living, industry standards, and market rates.
- Use data driven Insights: Recruiters should use data-driven insights to inform their rate discussions and ensure that salaries and compensation packages are competitive.

- Communicate effectively: Recruiters should communicate effectively with clients and candidates, ensuring that everyone is on the same page regarding salary expectations and cost-of-living considerations.

By considering the cost of living and factoring it into rate discussions and expectations, recruiters can ensure that clients and candidates have a clear understanding of the compensation package and its implications. This can lead to more effective recruitment and talent management strategies.

BILL RATES, PAY RATES AND MARGINS.

Discussing bill rates, pay rates, and margins can be a conversation, especially in industries that involve contracting or consulting. We will break it down and explore how to have these discussions with clarity.

Understanding the Basics

First of all, before diving into the Conversation or any conversation at all, it is crucial to understand the basics. Therefore, it is essential to understand the basics of bill rates, pay rates, and margins.

- Bill rate: The bill rate is the hourly or project-based rate that a client is charged for a contractor's or consultant's services.
- Pay rate: The pay rate refers to the hourly or salary-based compensation that a contractor or consultant receives for their work.
- Margin: The margin is the difference between the bill rate and the pay rate, which represents the profit or markup for the contracting or consulting company.

Discussing Bill Rates

When discussing bill rates, it is very necessary to be transparent and clear about the calculation.

- Explain the calculation: Explain how the bill rate is calculated, including any factors that may affect it, such as the contractor's experience, industry standards, or project complexity.
- Provide context: Provide context about the market rate for similar services and how the bill rate compares.

- Be open to negotiation: Be open to negotiating the bill rate, but also be clear about the minimum acceptable rate.

Discussing Pay Rates

When discussing pay rates, it is also necessary to be clear about the compensation structure.

- Explain the compensation structure: Explain how the pay rate is determined, including any factors that may affect it, such as performance, experience, or industry standards.
- Discuss benefits and perks: Discuss any benefits or perks that are included in the compensation package, such as health insurance, retirement plans, or paid time off.
- Be transparent about expectations: Be transparent about the expectations for the role and how the pay rate is tied to performance.

Discussing Margins

When discussing margins, it is crucial to be transparent about the calculation and the reasoning behind it.

- Explain the margin calculation: Explain how the margin is calculated and what factors affect it, such as overhead costs, profit targets, or industry standards.
- Provide context: Describe the industry standard for margins and compare the company's margin accordingly.
- Be transparent about pricing strategy: Be transparent about the pricing strategy and how it affects the margin.

How to have a Clear Communication

To ensure clear communication when discussing bill rates, pay rates, and margins, follow these best practices:

- Use simple language: Avoid using jargon or technical terms that may confuse the Conversation.
- Be transparent: Be transparent about the calculation and reasoning behind the bill rate, pay rate, and margin.

- Provide context: Provide context about industry standards, market rates, and other relevant factors.
- Be open to negotiation: Be open to negotiating the bill rate, pay rate, or margin, but also be clear about the minimum acceptable terms.

By following these practices and being transparent about bill rates, pay rates, and margins, you can have productive and transparent conversations with clients, contractors, or consultants.

PAY RATE vs BILL RATE vs MARKUP

Let's break it down:

- Pay Rate: This is the amount the Candidate earns for their work. It is the hourly or salary rate paid to the Candidate by the staffing agency or consultancy.
- Bill Rate: This is the amount the client pays for the Candidate's services. It is the hourly or project-based rate charged to the client by the staffing agency or consultancy.
- Markup: This is the difference between the bill rate and the pay rate. It represents the profit or margin earned by the staffing agency or consultancy. The markup is calculated by subtracting the pay rate from the bill rate.

For example, if the bill rate is $100 per hour and the pay rate is $70 per hour, the markup would be $30 per hour ($100 - $70). This means the staffing agency or consultancy earns a profit of $30 per hour for each hour the Candidate works.

DIAGRAM.

BILL RATE VS PAY RATE BREAKDOWN.

**BILL RATE VS PAY RATE
BREKDOWN FLOWCHART**

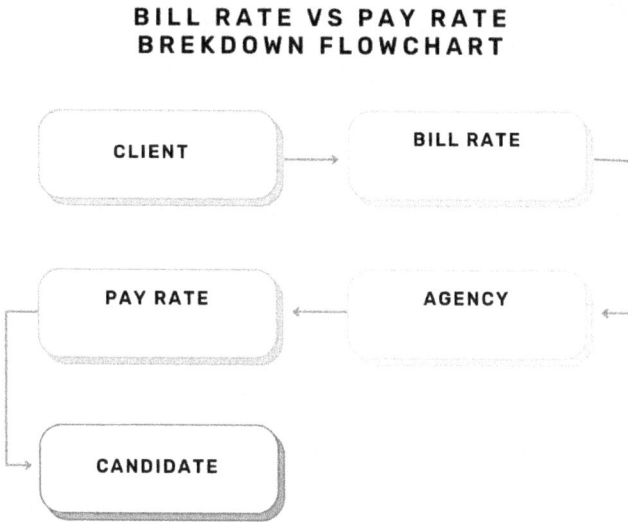

Note: Agency retains margin or markup

DIAGRAM SHOWING DISBURSEMENT OF CASH.

Name of Service: Disbursement of Cash

PROCESS FLOW CHART

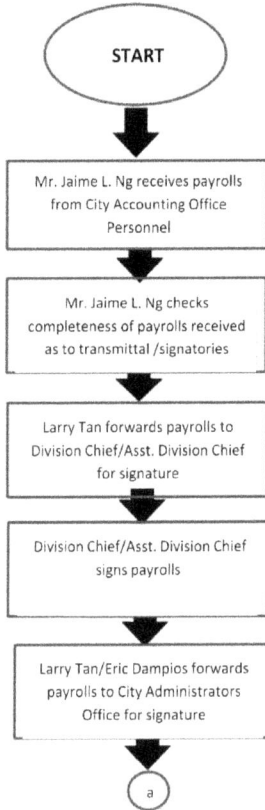

```
                    ┌─────────────┐
                    │    START    │
                    └─────────────┘
                           │
                           ▼
        ┌──────────────────────────────────┐
        │  Mr. Jaime L. Ng receives payrolls │
        │    from City Accounting Office     │
        │             Personnel              │
        └──────────────────────────────────┘
                           │
                           ▼
        ┌──────────────────────────────────┐
        │     Mr. Jaime L. Ng checks        │
        │  completeness of payrolls received │
        │   as to transmittal /signatories   │
        └──────────────────────────────────┘
                           │
                           ▼
        ┌──────────────────────────────────┐
        │   Larry Tan forwards payrolls to   │
        │ Division Chief/Asst. Division Chief │
        │           for signature            │
        └──────────────────────────────────┘
                           │
                           ▼
        ┌──────────────────────────────────┐
        │ Division Chief/Asst. Division Chief │
        │           signs payrolls           │
        └──────────────────────────────────┘
                           │
                           ▼
        ┌──────────────────────────────────┐
        │ Larry Tan/Eric Dampios forwards    │
        │   payrolls to City Administrators  │
        │        Office for signature        │
        └──────────────────────────────────┘
                           │
                           ▼
                          ( a )
```

NEGOTIATING COMPENSATION.

Negotiating compensation can be a challenging but crucial part of the recruitment process. Building confidence in negotiation skills can help ensure that client and candidate expectations are met. Here are some tips to build confidence and negotiate effectively:

1. Understand the Market

- Research industry standards: Understand the market rate for similar roles and industries within your field.

- Know the client's budget: Be aware of the client's budget and expectations.
- Know the Candidate's expectations: Understand the Candidate's salary requirements and expectations.

2. Prepare for the Conversation

- Gather data: Gather data on market rates, industry standards, and the client's budget.
- Anticipate concerns: Anticipate concerns or objections that may arise during the Conversation.
- Develop a negotiation strategy: Develop a negotiation strategy that takes into account the client's and Candidate's expectations.

3. Build Rapport and Trust

- Establish a relationship: Build a relationship with the client and Candidate based on trust and transparency.
- Communicate effectively: Communicate effectively and clearly about the negotiation process.
- Show empathy: Show empathy and understanding for the client's and candidate's needs and concerns.

4. Negotiate with Confidence

- Be confident but respectful: Be confident in your negotiation skills but also respectful of the client's and candidate's needs.
- Use data to support your position: Use data to support your position and negotiate effectively.
- Be flexible: Be flexible and open to creative solutions that meet the client's and candidate's needs.

5. Close the Deal

- Summarise the agreement: Summarise the agreement to ensure all parties are on the same page.
- Confirm the details: Confirm the details of the agreement, including the compensation package.
- Follow up: Follow up with the client and Candidate to ensure that the agreement is implemented correctly.

By following these tips, you can build confidence in your negotiation skills and negotiate compensation that meets client and candidate expectations.

NEGOTIATION TIPS FOR GLOBAL RECRUITERS.

Be Transparent

Honesty is key: Being transparent about what you know and what you don't know builds trust with clients and candidates. It is essential to be upfront about the recruitment process, timelines, and any potential challenges.

- Builds credibility: Transparency helps establish credibility and demonstrates expertise in the recruitment process.

Focus on Value

- Highlight the benefits: Instead of just focusing on numbers, highlight the value that the Candidate brings to the client's organization. This could include skills, experience, or specific achievements that will benefit the client.
- Emphasize the ROI: Explain how the Candidate's skills and experience will generate a return on investment (ROI) for the client. This could include cost savings, increased productivity, or improved efficiency.

Confirm the Model

- Understand client preferences: Early in the intake process, confirm what model works best for the client. This could include W2, C2C, or other engagement models.
- Ask the right questions: Ask questions like "Do you prefer W2, C2C, or open to both?" to understand the client's preferences and tailor your approach accordingly.

Additional Tips

- Active listening: Pay attention to the client's needs and concerns and respond accordingly.
- Flexibility: Be open to creative solutions and flexible in your approach to meet the client's needs.
- Clear communication: Ensure that all communication is clear, concise, and transparent.

By following these negotiation tips, global recruiters can build strong relationships with clients, secure top talent, and drive successful recruitment outcomes.

Understanding compensation is crucial in U.S. IT recruitment due to the complexity of pay models, legal structures, and financial expectations. Here's why:

Multiple Pay Models

- Different pay structures: U.S. IT recruitment involves various pay models, including hourly, salaried, W-2, 1099, and C2C. Each model has its implications for clients and candidates.
- Client and candidate expectations: Understanding these pay models ensures that client budgets and candidate expectations are aligned.

Legal Structures

- Compliance: Familiarity with legal structures, such as W-2, 1099, and C2C, is essential to ensure compliance with U.S. laws and regulations.
- Implications for clients and candidates: Each legal structure has different implications for clients and candidates, including tax obligations, benefits, and liability.

Financial Expectations

- Candidate value: Understanding compensation ensures that candidates are fairly compensated for their skills and experience.
- Client budget: It also ensures that clients' budgets are protected and their financial expectations are met.

Consequences of Misalignment

- Rejections and counteroffers: Misalignment between client budgets and candidate expectations can lead to rejections and counteroffers, which can delay the recruitment process and damage relationships.
- Loss of top talent: Failure to understand compensation can result in the loss of top talent to competitors who offer more competitive compensation packages.

Benefits of Understanding Compensation

- Successful placements: Understanding compensation ensures that placements meet the needs of both clients and candidates, leading to successful outcomes.
- Stronger relationships: It helps build stronger relationships with clients and candidates, leading to repeat business and referrals.
- Competitive advantage: Recruiters who understand compensation can differentiate themselves from competitors and provide more value to clients and candidates.

In summary, understanding compensation is crucial in U.S. IT recruitment to ensure successful placements, foster stronger relationships, and maintain a competitive edge.

Case Example: A recruiter placed a DevOps contractor at $80/hr on C2C. The client's budget was $95 per hour. Instead of maxing the margin, the recruiter offered $85/hr to the Candidate and retained a smaller margin—but gained the client's long-term trust and the Candidate's loyalty.

This case example highlights the importance of striking a balance between maximizing margins and fostering long-term relationships in recruitment. Here's a breakdown of the scenario:

The Situation

- Client budget: The client had a budget of $95 per hour for the DevOps contractor.
- Recruiter's decision: Instead of maximizing the margin, the recruiter decided to offer the Candidate $85/hour, retaining a smaller margin.

The Outcome

- Client trust: By not exceeding the client's budget, the recruiter demonstrated transparency and fairness, thereby gaining the client's long-term trust.
- Candidate loyalty: The Candidate received a competitive rate of $85/hour, which likely led to increased job satisfaction and loyalty.

The Benefits

- Long-term partnership: The recruiter's decision helped build a long-term partnership with the client, potentially leading to future business opportunities.
- Positive reputation: The recruiter's transparency and fairness may have enhanced their reputation in the industry, attracting more clients and candidates.
- Candidate retention: The Candidate's loyalty may lead to longer tenure and reduced turnover costs for the client.

The Lesson

- Balance margins with relationships: Recruiters should strike a balance between their margins and building long-term relationships with clients and candidates.
- Transparency and fairness: Demonstrating transparency and fairness can lead to increased trust and loyalty from both clients and candidates.

NOTE: This example is scenario-based training and learning for our purposes to understand the scope better. *** The scope can be much deeper and more precise.

Self-Assessment #6

1. What is the primary difference between W2 and C2C?
2. What does 1099 mean in the context of U.S. recruitment?
3. Why is it important to understand the cost of living when discussing salary?
4. What is the difference between bill rate and pay rate?
5. How can a recruiter's negotiation skills affect placement success?

Reflective Prompt: Think of a time you discussed or negotiated a candidate's rate. What strategy worked well? What would you improve next time?

CHAPTER SEVEN

ATS, CRM and VMS SYSTEMS

Objectives of this Chapter:

- Understand what an ATS (Applicant Tracking System) is and how it supports recruitment
- Learn the purpose and structure of VMS (Vendor Management Systems)
- Explore CRM (Candidate Relationship Management) platforms and their strategic role
- Identify key differences and integrations among ATS, VMS, and CRM tools
- Develop efficiency and organization in your recruitment workflow through systems knowledge

An Applicant Tracking System (ATS) is a software application that enables companies to electronically manage their job postings, candidate applications, and resumes. In this chapter, we will discuss how it supports recruitment.

What is an ATS?

- Automated recruitment process: An ATS automates the recruitment process, allowing companies to manage job postings, candidate applications, and resumes in a centralized system.
- Streamlined workflow: ATSs provide a streamlined workflow for recruiters and hiring managers to track candidate progress, schedule interviews, and make hiring decisions.

Benefits of ATS

- Increased efficiency: ATSs save time and effort for recruiters and hiring managers by automating tasks such as resume screening and candidate sorting.
- Improved candidate experience: ATSs provide a user-friendly interface for candidates to apply for jobs and track their application status.

- Better data management: ATSs enable companies to store and manage candidate data in a centralized system, making it easier to track candidate pipelines and analyze recruitment metrics effectively.ATSs provide insights into recruitment metrics, such as time-to-hire, source of hire, and candidate pipelines.
- Enhanced collaboration: ATSs enable recruiters and hiring managers to collaborate on candidate selection and communication.

Key Features of an ATS

- Job posting: ATSs enable companies to post job openings on their websites, social media platforms, and job boards.
- Resume parsing: ATSs extract relevant information from resumes, making it easier to search and filter candidates.
- Candidate tracking: ATSs track candidate applications, resumes, and communication, providing a centralized database for recruiters and hiring managers to manage their recruitment processes effectively.
- Automated screening: ATSs can automatically screen candidates based on predefined criteria, such as specific keywords, relevant experience, and required qualifications.
- Interview scheduling: ATSs often include tools for scheduling interviews and sending automated notifications to candidates.

Challenges of ATS

- Keyword optimization: Candidates need to optimize their resumes with relevant keywords to pass through ATS filters.
- Limited visibility: Some ATSs may limit visibility into the recruitment process, making it difficult for candidates to track their application status.
- Bias and fairness: ATSs can perpetuate biases if not correctly configured, potentially leading to unfair treatment of certain candidate groups.

Impact on Recruitment

- Increased competition: ATSs can intensify competition for job openings, as candidates must differentiate themselves in a digital applicant pool.
- Changes in recruiter role: ATSs require recruiters to adapt their role, focusing more on strategic tasks such as candidate sourcing and relationship building.

- Candidate experience: ATSs can impact the candidate experience, either positively or negatively, depending on the system's usability and configuration.

Overall, ATSs have transformed the recruitment landscape, offering both benefits and challenges for companies, recruiters, and candidates. By understanding how ATSs work and adapting to their requirements, recruiters can optimize their recruitment strategies and enhance the candidate experience.

How an ATS Supports Recruitment

- Streamlines the hiring process: ATSs simplifies the application and screening process, making it easier to manage large volumes of candidates.
- Improves candidate quality: ATSs help recruiters identify top candidates by automating screening and filtering based on predefined criteria.
- Enhances candidate engagement: ATSs provide a positive candidate experience through timely communication and updates on application status.
- Supports data-driven decision-making: ATSs provide valuable insights into recruitment metrics, enabling informed decision-making and process improvements.

By automating and streamlining the recruitment process, ATSs help companies manage their hiring needs more efficiently and effectively.

KEYWORD OPTIMIZATION.

Under the challenges of ATS, we mentioned keyword optimization, though it wasn't explained. And that is what we will be doing here.

What is keyword optimization?

Keyword optimization is the process of incorporating relevant words and phrases into a resume to increase its visibility and chances of passing through Applicant Tracking System (ATS) filters. This involves using keywords and phrases from the job description and requirements to match the ATS's keyword-matching criteria, ultimately improving the Candidate's chances of being seen by recruiters and hiring managers.

- ATS filtering: Many companies use Applicant Tracking Systems (ATSs) to filter and screen resumes before they even reach a human recruiter or hiring manager.
- Keyword matching: ATSs often use keyword matching to identify relevant candidates. This means the system looks for specific words or phrases in a resume that match the job requirements.

Why is keyword optimization important?

- Passing the ATS filter: If a resume doesn't contain the right keywords, it may not pass through the ATS filter, and the Candidate may not be considered for the job.
- Increasing visibility: By including relevant keywords, candidates can enhance the visibility of their resume and increase their chances of being noticed by recruiters or hiring managers.

How to optimize your resume with keywords

- Read the job description: Carefully read the job description and requirements to identify keywords and phrases.
- Use relevant keywords: Incorporate relevant keywords into your resume, especially in sections such as:
- Summary/Objective
- Skills
- Work experience
- Use keyword variations: Use variations of keywords to avoid repetition and increase the chances of matching the ATS filter.
- Be specific: Use specific keywords and phrases rather than general terms.

Examples of keywords

- Technical skills: Programming languages (e.g., Java, Python), software proficiency (e.g., Adobe Creative Suite), or technical certifications (e.g., AWS Certified Developer).
- Industry-specific terms: Terms related to a specific industry, such as healthcare (e.g., HIPAA, EMR) or finance (e.g., financial modelling, forecasting).
- Soft skills: Skills like communication, teamwork, or leadership.

Understanding Vendor Management Systems (VMS)

In today's fast-paced business landscape, companies are increasingly relying on contingent workers, freelancers, and temporary staff to meet their talent needs. Managing these workers can be quite a difficult task, which is where Vendor Management Systems (VMS) come into play.

DIAGRAM.

ATS WORKFLOW FROM JOB REQUISITION TO PLACEMENT.

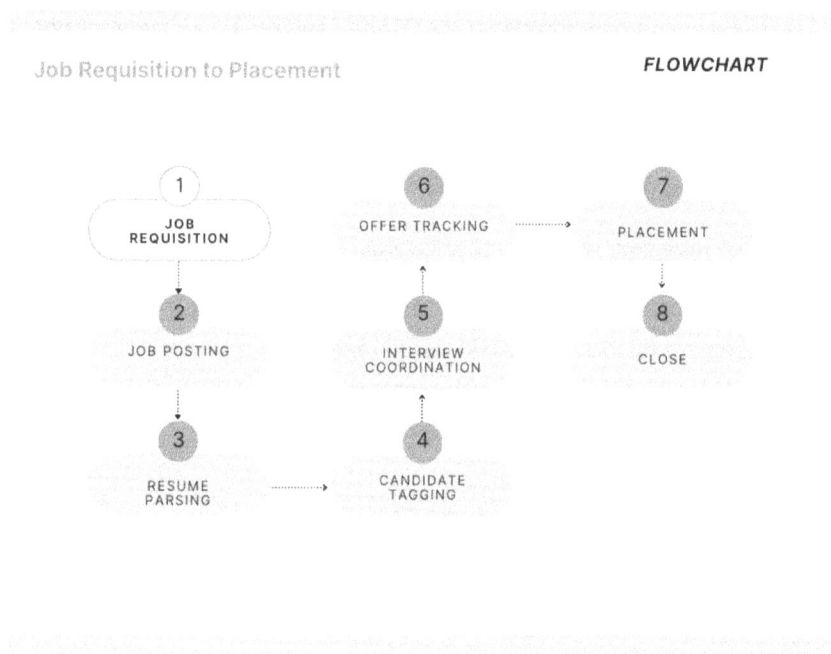

Job Requisition to Placement **FLOWCHART**

1. JOB REQUISITION
2. JOB POSTING
3. RESUME PARSING
4. CANDIDATE TAGGING
5. INTERVIEW COORDINATION
6. OFFER TRACKING
7. PLACEMENT
8. CLOSE

What is a VMS?

A Vendor Management System (VMS) is a software application that enables companies to manage their contingent workforce, including temporary workers, freelancers, and contract employees. A VMS provides a centralized platform for managing the entire contingent workforce lifecycle, from sourcing and recruitment to onboarding, time tracking, and payment.

Purpose of a VMS

The primary purpose of a VMS is to streamline and optimize the management of contingent workers. By automating many of the manual processes involved in managing temporary staff, a VMS helps companies to:

- Reduce costs: By automating processes and improving visibility into contingent workforce spending, companies can better manage their budgets and reduce costs.
- Improve efficiency: A VMS helps companies streamline their recruitment processes, reduce time-to-hire, and enhance the overall quality of contingent workers.

Enhance compliance: A VMS can help companies ensure compliance with labour laws and regulations, thereby reducing the risk of non-compliance and associated penalties.

- Gain visibility: A VMS provides real-time visibility into contingent workforce spending, enabling companies to make informed decisions about their talent needs.

Structure of a VMS

A typical VMS consists of several key components, including:

- Sourcing and recruitment: A VMS provides tools for sourcing and recruiting contingent workers, including job posting, candidate tracking, and resume management.
- Onboarding and compliance: A VMS helps companies to onboard contingent workers efficiently, ensuring that all necessary paperwork and compliance requirements are met.
- Time tracking and payment: A VMS provides tools for tracking the time worked by contingent workers, enabling companies to manage payment and invoicing processes effectively.
- Reporting and analytics: A VMS provides real-time reporting and analytics, enabling companies to gain insights into contingent workforce spending, productivity, and performance.

- Vendor management: A VMS enables companies to effectively manage their relationships with staffing suppliers and other vendors, ensuring they receive the best talent at the most competitive prices.

Benefits of a VMS

The benefits of a VMS are numerous, including:

- Cost savings: A VMS can help companies reduce their contingent workforce costs by automating processes and improving visibility into spending.
- Improved efficiency: A VMS streamlines recruitment processes, reducing time-to-hire and improving the overall quality of contingent workers.
- Enhanced compliance: A VMS helps companies to ensure compliance with labour laws and regulations, reducing the risk of non-compliance and associated penalties.
- Better decision-making: A VMS provides real-time visibility into contingent workforce spending, enabling companies to make informed decisions about their talent needs.
- Enhanced visibility: A VMS provides real-time visibility into contingent labour spend, vendor performance, and other key metrics.

Key Features of a VMS

1. Candidate submission and tracking: A VMS allows companies to submit candidate requests to multiple staffing agencies and track the status of those requests.
2. Feedback management: A VMS enables companies to provide feedback to staffing agencies and track the performance of candidates.
3. Compliance management: A VMS helps companies ensure compliance with labour laws and regulations, including worker classification and benefits.
4. Time and expense tracking: A VMS allows companies to track billable hours and expenses for contingent workers.
5. Reporting and analytics: A VMS provides insights into contingent labour spend, vendor performance, and other key metrics.

Common VMS Platforms

Some popular VMS platforms include:

1. Beeline: A cloud-based VMS that provides a comprehensive platform for managing contingent labour.
2. Fieldglass (SAP): A VMS platform that provides a range of tools for managing contingent labour, including candidate sourcing and onboarding.
3. IQNavigator: A VMS platform that provides a range of tools for managing contingent labour, including candidate sourcing and time tracking.
4. PRO Unlimited: A VMS platform that provides a range of tools for managing contingent labour, including candidate sourcing and compliance management.

MSP Environments

VMS platforms are often utilized in Managed Service Provider (MSP) environments, where a third-party provider manages a company's contingent labour needs. MSPs use VMS platforms to manage multiple staffing agencies and vendors, ensuring that companies receive high-quality talent at competitive rates.

In conclusion, a Vendor Management System (VMS) is a powerful tool for managing contingent workers. By providing a centralized platform for managing the entire contingent workforce lifecycle, a VMS helps companies reduce costs, improve efficiency, enhance compliance, and gain visibility into their talent needs. Whether you're a small business or a large enterprise, a VMS can help you optimize your contingent workforce management processes and achieve your business goals.

CANDIDATE RELATIONSHIP MANAGEMENT (CRM)

What is Candidate Relationship Management (CRM)?

In the world of recruitment, building and maintaining relationships with potential candidates is crucial for success. Candidate Relationship Management (CRM) is a strategic approach that enables recruiters to nurture and manage long-term relationships with candidates, ultimately leading to improved hiring outcomes.

Key Features of a CRM System

A Candidate Relationship Management system is designed to help recruiters build and maintain relationships with candidates. Some key features of a CRM system include:

1. Candidate Profiling: A CRM system enables recruiters to create comprehensive profiles of candidates, encompassing their skills, experience, and preferences.
2. Conversation Tracking: Recruiters can track conversations with candidates, including emails, phone calls, and messages, to ensure timely followups and personalized communication.
3. Availability and Scheduling: A CRM system can help recruiters manage candidate availability and schedule interviews, assessments, or other interactions.
4. Preference Management: Recruiters can store candidate preferences, such as job roles, locations, or salary expectations, to tailor their communication and job recommendations.
5. Engagement Analytics: A CRM system provides insights into candidate engagement, enabling recruiters to measure the effectiveness of their strategies and make data-driven decisions.

Benefits of Using a CRM System

1. Improved Candidate Experience: A CRM system enables recruiters to provide personalized communication and tailored job recommendations, enhancing the candidate experience.
2. Increased Efficiency: By automating routine tasks and providing real-time insights, a CRM system helps recruiters streamline their workflow and focus on high-value activities.
3. Better Hiring Outcomes: A CRM system enables recruiters to build relationships with candidates, increasing the chances of attracting top talent and reducing the time to hire.
4. Data-Driven Decision Making: By analyzing candidate engagement and behaviour, recruiters can make informed decisions about their recruitment strategies.

Integration with ATS

Many CRM systems are integrated with Applicant Tracking Systems (ATS), allowing recruiters to manage candidate relationships and applications seamlessly. This integration provides a unified view of candidate interactions, streamlining the recruitment process and improving efficiency.

Popular CRM Platforms

Some popular CRM platforms for recruitment include:

1. Salesforce Talent: A cloud-based CRM platform that provides a comprehensive solution for managing candidate relationships.
2. Avionté: A recruitment CRM platform that offers advanced features for candidate engagement and workflow automation.
3. Beamery: A talent CRM platform that provides AI-powered insights and personalized candidate experiences.

Best Ways for Implementing a CRM System

1. Define Clear Goals: Establish clear goals and objectives for your CRM system, such as improving candidate engagement or reducing time to hire.
2. Customize Your Approach: Tailor your CRM system to meet your organization's specific needs and recruitment strategies.
3. Train Your Team: Provide comprehensive training to your recruitment team to ensure they can utilize the CRM system effectively.
4. Monitor and Analyse Performance: Regularly monitor and analyse performance metrics to optimise your recruitment strategies and enhance candidate relationships.

By implementing a CRM system and adhering to best practices for its implementation, recruiters can foster strong relationships with candidates, drive more effective hiring outcomes, and ultimately achieve their recruitment objectives.

KEY DIFFERENCES AND INTEGRATIONS AMONG ATS, VMS AND CRM TOOLS.

So, we are going to be talking about the differences between ATS, VMS and CRM tools. Before we go into it, we are going to define and explain them again briefly.

ATS (Applicant Tracking System), VMS (Vendor Management System), and CRM (Candidate Relationship Management) tools are essential components of modern recruitment processes. While they serve distinct purposes, they often integrate to streamline workflows and improve efficiency.

ATS (Applicant Tracking System)

- Purpose: Manage job postings, candidate applications, and hiring workflows.
- Key Features: Job posting, resume parsing, candidate screening, interview scheduling, and hiring decision-making.
- Benefits: Streamlines recruitment processes, reduces time to hire and improves candidate quality.
- VMS (Vendor Management System)
- Purpose: Manage relationships with staffing agencies, track vendor performance, and automate invoicing and payments.
- Key Features: Vendor onboarding, job posting, candidate submission, timesheet management, and invoicing.
- Benefits: Improves vendor management, reduces costs, and enhances compliance.

CRM (Candidate Relationship Management)

- Purpose: Build and maintain relationships with candidates, personalize communication, and improve candidate experience.
- Key Features: Candidate profiling, conversation tracking, availability and scheduling, and engagement analytics.
- Benefits: Enhances candidate experience, improves candidate quality, and increases retention.

Integrations among ATS, VMS, and CRM Tools

1. ATS + CRM: Integrates candidate application management with relationship-building features, enabling recruiters to personalize communication and improve candidate experience.
2. ATS + VMS: Integrates job posting and candidate management with vendor management, enabling staffing agencies to manage client relationships and candidate submissions efficiently.

3. CRM + VMS: Integrates candidate relationship management with vendor management, enabling recruiters to build relationships with candidates and manage vendor performance.

Benefits of this Integration:

1. Streamlined Workflows: Integrations eliminate manual data entry, reduce errors, and improve efficiency.
2. Enhanced Candidate Experience: Integrations enable recruiters to personalize communication, enhance candidate engagement, and foster stronger relationships.
3. Enhanced Data Insights: Integrations offer a unified view of candidate interactions, vendor performance, and recruitment metrics, facilitating data-driven decision-making.

By understanding the differences and integrations among ATS, VMS, and CRM tools, recruiters can optimise their recruitment processes, enhance the candidate experience, and achieve more effective hiring outcomes.

Benefits of Recruitment Systems in General.

Recruitment systems, such as ATS, VMS, and CRM, offer numerous benefits to organizations, including:

Staying Compliant

1. EEOC (Equal Employment Opportunity Commission) Compliance: Recruitment systems help ensure compliance with EEOC regulations by tracking and documenting hiring processes, reducing the risk of discriminatory practices.
2. OFCCP (Office of Federal Contract Compliance Programs) Compliance: These systems enable organizations to maintain records and reports required by OFCCP, ensuring compliance with federal contracting regulations.

Avoiding Duplication

1. Reducing Duplicate Candidates. Recruitment systems help identify and eliminate duplicate candidate submissions, reducing unnecessary work and improving efficiency.

2. Streamlining Candidate Management: By centralizing candidate data, these systems prevent duplication of effort and ensure that recruiters work with the most qualified candidates.

Building Candidate Pipelines

1. Candidate Relationship Management: Recruitment systems enable organizations to build and maintain relationships with candidates, creating a pipeline of qualified talent for future openings.
2. Talent Pool Management: These systems enable recruiters to identify, engage, and nurture top talent, thereby reducing the time to hire and improving candidate quality.

Reporting on Recruiter KPIs

1. Time-to-Hire: Recruitment systems provide valuable insights into time-to-hire metrics, enabling organisations to optimise their recruitment processes.
2. Source of Hire: These systems track the source of hire, enabling organizations to identify the most effective recruitment channels.
3. Recruiter Performance: Recruitment systems enable organizations to track recruiter performance, identifying areas for improvement and optimizing recruitment strategies.

Communicating Consistently

1. Candidate Communication: Recruitment systems enable organizations to communicate consistently with candidates, improving the candidate experience and reducing confusion.
2. Client Communication: These systems facilitate consistent communication with clients, ensuring that stakeholders are informed and engaged throughout the recruitment process.

Case Example: Benefits of ATS-VMS Integration Training

A recruiter's mistake in submitting the same Candidate twice to the VMS (once through ATS and once via email) resulted in a duplication violation, causing her to miss out on a high-paying role. However, after receiving proper training on her ATS-VMS integration, she was able to:

1. Avoid duplicate submissions: The recruiter learned how to manage candidate submissions properly, eliminating duplication errors.
2. Improve turnaround time: With a deeper understanding of the ATS-VMS integration, the recruiter was able to streamline her workflow, resulting in a 40% reduction in turnaround time.

Key Takeaways

1. Proper training is essential: Adequate training on ATS-VMS integration can help recruiters avoid common mistakes and improve efficiency.
2. Integration knowledge is crucial: Understanding how to leverage ATS-VMS integration can help recruiters work more efficiently and effectively.
3. Improved productivity: By avoiding duplication errors and streamlining workflows, recruiters can focus on high-value tasks, such as building relationships with candidates and clients.

(Tech Point Africa), (Paul Mitchen On LinkedIn)

This case example highlights the importance of proper training and integration knowledge in recruitment systems, enabling recruiters to work more efficiently and effectively.

NOTE: This example is scenario-based for training and learning purposes to help us understand the scope better. The real scope can be much deeper and more precise.

Self-Assessment.

1. What is the primary function of an ATS?
2. How does a VMS differ from an ATS?
3. Name one benefit of using a CRM as part of your recruitment process.
4. Why is system compliance substantial in global recruitment?
5. What's the relationship between a recruiter's ATS and a client's VMS?

Reflective Prompt: Consider the systems your current recruitment process employs. Are they helping you stay organized and efficient—or do they feel like extra work? What can you do to integrate your tools better and improve performance?

CHAPTER EIGHT

BOOLEAN STRINGS, JOB BOARDS & SOURCING STRATEGIES

Boolean Logic in Recruitment.

Boolean logic is a powerful tool used in recruitment to refine search queries and identify top talents. By using Boolean operators, recruiters can create targeted searches that yield more relevant results.

It is used particularly when searching for candidates online. It enables recruiters to refine their search queries using operators such as "AND," "OR," and "NOT." Here is how it applies:

AND: Used to find candidates with multiple specific skills or qualifications. For example, "software engineer AND Java AND Python" would yield results with all three terms.

OR: Expands search results to include candidates with at least one of the specified skills. For instance, "marketing OR sales OR business development" would return results with any of these terms.

NOTE: Excludes specific terms from search results. For example, "software engineer NOT junior" would return results without the term "junior."

Quotation marks: Used to search for exact phrases. For example, "digital marketing specialist" would return results with this exact phrase.

Parentheses: Groups search terms to clarify the order of operations. For instance, "(software engineer OR developer) AND Java" would return results with either "software engineer" or "developer" and "Java."

By applying Boolean logic, recruiters can:

- Streamline candidate searches
- Increase search precision
- Reduce time spent reviewing irrelevant results

- Find top talent more efficiently

APPLICATIONS OF BOOLEAN LOGIC IN RECRUITMENT

1. Resume Search: Boolean logic enables recruiters to search for resumes based on specific skills, experience, and qualifications.
2. Candidate Sourcing: Boolean searches can be utilized to identify potential candidates on job boards, social media platforms, and professional networking sites.
3. Applicant tracking system (ATS): Many ATS platforms support Boolean search functionality, enabling recruiters to refine their searches and identify top candidates.

BENEFITS OF BOOLEAN LOGIC IN RECRUITMENT

1. Improved search results: Boolean logic helps recruiters create targeted searches that yield more relevant results.
2. Increased efficiency: By refining search inquiries, recruiters can save time and focus on high-value high-value tasks.
3. Better candidate quality: Boolean logic enables recruiters to identify candidates with specific skills and experience, improving the quality of hire.

BUILDING BOOLEAN SEARCH STRINGS FOR IT ROLES.

Boolean search strings are a powerful tool for recruiters and talent acquisition professionals to find the right candidates for specific IT roles. By combining keywords, operators, and syntax, you can create targeted searches that yield relevant results. In this chapter, we will be talking about how to build boolean search strings for IT roles;

SOFTWARE ENGINEER.

- Search string; [software engineer OR developer] AND [Java OR python OR c++] AND [senior OR lead]
- Explanation: This search string looks for candidates with experience as a software engineer or developer, proficient in Java, Python, or C++, and with senior or lead-level experience.

CLOUD ARCHITECT

- Search string; 'Cloud architect' AND (AWS OR Azure OR Google Cloud) AND (design OR implementation OR migration)
- Explanation: This search string targets candidates with experience as a cloud architect, specifically in AWS, Azure, or Google Cloud, and with expertise in design, implementation, or migration.

CYBERSECURITY SPECIALIST

- Search string: (cybersecurity OR security) AND (analyst OR specialist OR engineer) AND (compliance OR risk management OR threat analysis)
- Explanation: This search string searches for candidates with experience as a data scientist or data analyst, with expertise in machine learning, deep learning, or predictive analytics, and proficiency in Python, R, or SQL.

TIPS AND BEST WAYS TO BUILD BOOLEAN STRINGS.

1. Use specific words; Use relevant keywords and phrases particular to the IT role you are searching for.
2. Combine operators: Use a combination of Boolean operators to refine your search results.
3. Use parentheses; Group search terms using parentheses to ensure the correct order of operations.
4. To rest and refine, Test your search strings and refine them based on the results.
5. Keep it concise: Keep your search strings concise and focused to avoid overwhelming the search engine.

COMMON CHALLENGES AND SOLUTIONS.

1. Too many results; Refine your search using more specific keywords or keywords or operators.
2. Too few results: Broaden your search string using OR operators or related terms.
3. Irrelevant results: Use NOT operators to exclude specific terms or phrases.

TOP JOB BOARDS AND PLATFORMS FOR SOURCING US CANDIDATES.

Here are some of the best platforms to find top IT talent in the US;

1. RESUME DATABASES;

- SignalHire: Fetches verified candidate emails and offers powerful search filters. It integrates with ATS systems like RecruiterFlow and features a Chrome extension for easy candidate sourcing (starts at $0.4, with a 4.4-star rating).
- HireEz: Uses AI to discover qualified candidates on social media and career sites like LinkedIn and Indeed. It offers advanced search features and integration with Recruiterflow (custom pricing, 4.5-star rating).
- Seek out: Sources candidates from 800M+ public profiles, offering AI-powered matching and integration with RecruiterFlow (custom pricing, 4.5-star rating).

2. JOB POSTING SITES;

1. Indeed: A free job search tool with sponsored job posting options and integration with Recruiterflow (starts at $0, 4.3-star rating).
2. ZipRecruiter: Post jobs to 100+ job sites, offering candidate management and rating features (starts at $0, 4.8 rating).
3. Dice: A platform specifically focused on tech with AI-powered matching and integration with Recruiterflow (custom pricing, 4.3-star rating).

3. SOCIAL MEDIA AND NETWORKING SITES

- LinkedIn: Offers advanced search filters, job posting, and recruiting tools (starts at $170/month, 4.5-star rating).
- GitHub: A community-driven platform for tech talent, allowing boolean searches and filtering by location and skills (starts at $0, 4.7-star rating).
- Wellfound: networking site for candidates who are ready to start offering AI-AI-powered arch and integration with team tools (it starting $0, 4.44.4-star rating

4. PORTFOLIO PLATFORMS;

- Behance: A platform for creative professionals, showcasing portfolios and offering hire pages and messaging features (custom pricing, 4.5-star rating)/
- Carbonmade: A portfolio website featuring creative candidates with simple search and filtering options (custom pricing, 4.2-star rating).

  Other options include;

  Manatal: An all-in-one recruitment platform offering candidate sourcing, applicant tracking, and recruitment CRM capabilities, suitable for US-based recruiters (offers a 14-day free trial).

  CV Library and Reed are also notable mentions; although they are more focused on the UK job market, they can still help source candidates with a global presence.

CREATING REPEATABLE AND ADAPTABLE SOURCING STRATEGIES:

To develop effective sourcing strategies, consider the following steps:

1. Define Sourcing Goals

Identify hiring needs: Determine the roles, skills, and qualifications required for the position.

- Set clear objectives: Establish metrics for success, such as time to hire, candidate quality, or diversity.

2. Analyze Past Sourcing Efforts

- Review previous campaigns: Identify what worked and what didn't.

Gather data: Track key metrics, including the source of hire, candidate quality, and time to hire.

3. Develop a Sourcing Framework

- Create a sourcing plan: Outline the steps, tools, and resources needed.

Identify sourcing channels: Determine the most effective job boards, social media platforms, and networking sites.

4. Build a Candidate Persona

- Define ideal candidate characteristics: Identify skills, experience, and qualifications.
- Create a candidate profile: Outline demographics, interests, and behaviors.

5. Utilize Technology and Tools

- Applicant Tracking Systems (ATS): Leverage ATS features, such as resume parsing and candidate tracking.
- Sourcing software: Utilize tools, such as Boolean search and candidate databases.

6. Continuously Monitor and Adapt

- Track metrics: Monitor time to hire, candidate quality, and other key performance indicators.
- Adjust strategies: Refine sourcing approaches based on data and feedback.

Best Practices

1. Stay up to date with industry trends: Continuously educate yourself on the latest sourcing techniques and tools to remain competitive.
2. Collaborate with hiring managers: Ensure alignment with hiring needs and objectives to maximize efficiency and effectiveness.
3. Diversify sourcing channels: Utilize multiple channels to reach a broader candidate pool.

By following these steps and best practices, you can create repeatable and adaptable sourcing strategies that drive results and improve hiring outcomes.

Improving Your Ability to Generate Targeted Candidate Pools Quickly

To generate targeted candidate pools efficiently, consider the following strategies:

Develop a Deep Understanding of the Job Requirements

1. Study the job description: Identify key skills, qualifications, and experience required.

2. Understand the hiring manager's needs: Clarify expectations and priorities to ensure alignment.

Make use of the Boolean Search Techniques

1. Master Boolean operators: Use AND, OR, NOT, and parentheses to refine searches.
2. Utilize search strings: Create targeted search strings using relevant keywords and phrases.

Utilize Advanced Search Features

1. LinkedIn Recruiter: Make use of advanced search filters, such as skills, experience, and education.
2. Other job boards and platforms: Utilize advanced search features on platforms like Indeed, Glassdoor, or GitHub.

Tap into Niche Communities

1. Industry-specific forums: Engage with online communities related to the job or industry.
2. Professional associations: Utilize networks and events to connect with potential candidates.

Leverage Employee Referrals

1. Employee referral programs: Encourage current employees to refer qualified candidates.
2. Incentivize referrals: Offer rewards or recognition for successful referrals.

Utilize AI Powered Sourcing Tools

1. AI-driven candidate sourcing: Leverage tools that use machine learning to identify top candidates.
2. Predictive analytics: Utilize tools that analyze data to predict candidate fit and potential.

Continuously Refine Your Approach

1. Track metrics: Monitor time to hire, candidate quality, and other key performance indicators.

2. Adjust strategies: Refine sourcing approaches based on data and feedback.

By implementing these strategies, you can enhance your ability to generate targeted candidate pools and achieve better hiring outcomes more quickly.

Why Boolean and Sourcing Matter

In today's competitive job market, relying solely on traditional recruitment methods, such as mass email blasts or job postings, is no longer sufficient. Proactive sourcing, driven by Boolean logic and strategic search techniques, has become increasingly crucial for identifying top talent, particularly in competitive domains such as cloud computing, security, and SaaS.

Benefits of Boolean and Sourcing

1. Precision: Boolean search enables recruiters to craft precise search strings, thereby reducing noise and enhancing the quality of candidates.
2. Efficiency: By targeting specific candidates, recruiters can save time and focus on high-value tasks.
3. Competitive Advantages: Proactive sourcing enables recruiters to identify top talent before competitors do, giving them a competitive edge.
4. Improved Candidate Quality: By targeting specific skills, experience, and qualifications, recruiters can attract higher-quality candidates.

Importance in Competitive Domains

1. Cloud: With the growing demand for cloud professionals, Boolean search enables recruiters to identify candidates with specific cloud skills, such as those in AWS or Azure.
2. Security: In the security domain, Boolean search enables recruiters to find candidates with specialized skills, such as penetration testing or security analysis.
3. SaaS: For SaaS roles, Boolean search helps recruiters identify candidates with experience in specific software or technologies.

How you can achieve this:

1. Master Boolean logic: Develop expertise in Boolean search techniques to craft effective search strings.
2. Understand the job requirements: Clearly define the skills, experience, and qualifications required for the role.
3. Utilize the right tools: Leverage sourcing tools, such as LinkedIn Recruiter or Boolean search software, to streamline the sourcing process.

By embracing Boolean and sourcing techniques, recruiters can efficiently and effectively deliver great candidates, even in competitive domains.

Boolean String Examples for Common Roles

Earlier in this chapter, we talked about a few roles for Boolean search strings, but right now, we will be talking about a few more. Here are some examples of Boolean search strings for typical roles:

1. Java Developer

- Search String: `(Java OR "Java SE") AND (Spring OR Hibernate) AND ("REST API") AND (Microservices)`
- Explanation: This search string looks for candidates with experience in Java, specifically with Spring or Hibernate frameworks, REST API development, and microservices architecture.

2. DevOps Engineer

- Search String: `("AWS Certified" OR "Azure DevOps") AND (Terraform OR CloudFormation) AND CI/CD`
- Explanation: This search string searches for candidates with experience in cloud platforms (AWS or Azure), infrastructure as code (Terraform or CloudFormation), and continuous integration/continuous deployment (CI/CD).

3. QA Automation

- Search String: `(Selenium OR Cypress) AND (Java OR Python) AND ("Test Automation")`

- Explanation: This search string looks for candidates with experience in test automation tools (Selenium or Cypress), programming languages (Java or Python), and test automation frameworks.

DIAGRAM

BOOLEAN TREE DIAGRAM

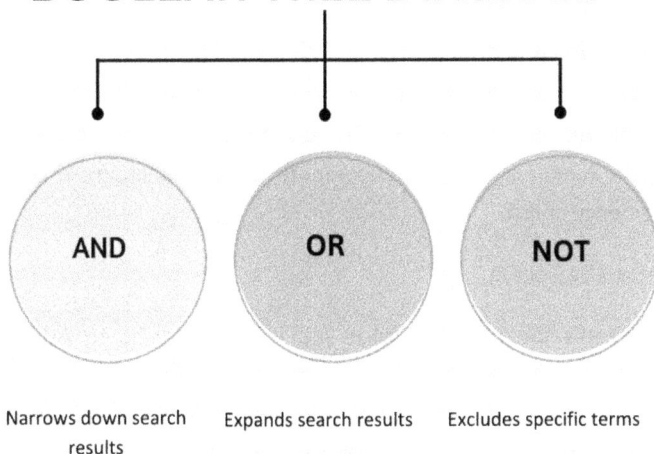

AND	OR	NOT
Narrows down search results	Expands search results	Excludes specific terms

By combining these techniques, you can efficiently narrow or expand candidate searches to find relevant results.

Global Tools and Platforms for Recruitment

Here are some global tools and platforms that can help you find top talent from around the world:

1. LinkedIn Recruiter

- Global visibility: LinkedIn Recruiter provides access to a vast pool of professionals worldwide.

- Advanced search filters: Use filters like location, industry, and skills to find the right candidates.

2. GitHub and Stack Overflow

- Developer communities: GitHub and Stack Overflow are popular platforms for developers to showcase their work and share knowledge with others.
- Identify top talent: Search for candidates with specific skills and experience in open-source projects or technical discussions.

3. HackerRank and Kaggle

- Data/AI communities: HackerRank and Kaggle are platforms for data scientists and AI professionals to showcase their skills.
- Competitions and challenges: Use these platforms to find candidates who have participated in competitions or challenges related to data science and AI.

4. Job Portals by Region

- Oceania job boards: Utilize job boards specific to Australia, New Zealand, or other Oceania countries to find local talent.
- Asia job boards: Use job boards popular in Asia, such as China, India, or Japan, to find candidates with specific skills and experience.
- Africa job boards: Leverage job boards specific to Africa to find talent in regions like South Africa, Nigeria, or Egypt.

Benefits of Using Global Tools and Platforms

1. Access to global talent: These platforms provide access to a vast pool of candidates from around the world.
2. Increased diversity: By sourcing candidates from different regions, you can increase diversity in your hiring process.
3. Improved candidate quality: These platforms enable you to find candidates with specific skills and experience, thereby enhancing the quality of your hires.

By utilizing these global tools and platforms, you can streamline your recruitment process and find top talent from around the world.

Sourcing Strategy Framework

Here is a step-by-step framework for developing an effective sourcing strategy:

1. Understand the Role's Domain/Sub-Domain

- Research the industry: Gain a deep understanding of the role's domain and subdomain.
- Identify key skills: Determine the essential skills, qualifications, and experience required for the role.

2. Write Boolean Strings

- Customize for platforms: Write 2-3 Boolean strings tailored to different platforms, such as LinkedIn, GitHub, or job boards.
- Use relevant keywords: Incorporate relevant keywords, phrases, and operators to refine your search results.

3. Choose Your Sources

- Job boards: Utilize job boards, such as Indeed, Glassdoor, or niche job boards.
- Databases: Make use of databases like LinkedIn Recruiter or candidate databases.
- Referrals: Encourage employee referrals or utilize professional networks to expand your reach.

4. Apply Filters

- Location: Filter candidates by location to ensure proximity to the workplace.
- Visa: Be aware of visa requirements and restrictions for international candidates.
- Experience: Filter candidates by experience level to match the role's requirements.

5. Save Strings/Templates

- Reuse and adjust: Save Boolean strings and templates for future use, adjusting them as needed to meet changing requirements.

- Refine your approach: Continuously refine your sourcing strategy based on the results and feedback you receive.

By following this framework, you can develop an effective sourcing strategy to identify top talent that meets your organization's needs.

Case Example: Finding React Developers in the US Southeast

A recruiter faced challenges finding React Developers in the US Southeast. To overcome this, he did this:

1. Adjusted the Boolean string: Included job title synonyms like "Frontend Engineer" and "UI Developer" to broaden the search scope.
2. Expanded the radius filter: Increased the geographic search area to include more potential candidates.

Result of the changes.

By making those necessary changes and adjustments, the recruiter was able to:

1. Source five candidates: Found five qualified React Developers in the US Southeast.
2. Achieve results quickly: Sourced candidates within two hours, demonstrating the efficiency of the adjusted search strategy.

Key Points to note:

1. Boolean string optimization: Using relevant synonyms and keywords can significantly improve search results.
2. Geographic flexibility: Expanding the search radius can help find candidates who may not be directly in the desired location.
3. Efficient sourcing: By refining the search strategy, recruiters can quickly find qualified candidates, saving time and effort.

This case example highlights the importance of adapting and refining search strategies to achieve successful recruitment outcomes.

NOTE: This example is scenario-based training and learning, with purposes that help further define the scope. ***The scope can be much deeper and more precise.

Self-Assessment

1. What does the AND operator do in a Boolean string?
2. Write a Boolean string for a data engineer with AWS and Python.
3. Which platform is best for finding passive US IT candidates?
4. Name two regional/global platforms aside from job boards for technical sourcing.
5. Why is it essential to customize your Boolean string per platform?

Reflective Prompt: Reflect on your most recent sourcing effort. What Boolean operators did you use? How could you improve that string to make your search more efficient next time?

CHAPTER NINE

GLOBAL SOURCING AND IMMIGRATION CONSIDERATIONS

Objectives of this chapter:

- Understand when and how to source global candidates for U.S.-based roles
- Learn the basics of U.S. work authorization types (H1B, OPT, EAD, Green Card)
- Identify risks and opportunities with visa-dependent candidates
- Know how to evaluate immigration eligibility during screening
- Use ethical and informed practices when sourcing internationally.

Sourcing Global Candidates for U.S.Based Roles

Sourcing global candidates for U.S.-based roles has become a crucial strategy for companies seeking to tap into a diverse pool of talent. However, it requires careful planning, execution, and consideration of various factors. In this chapter, we will give you a detailed guide on when and how to source global candidates for U.S.-based roles.

When to Source Global Candidates

1. Skill shortages: When specific skills are in short supply in the US market, sourcing global candidates can help fill the gap.
2. Diversity and inclusion: Global sourcing can bring diverse perspectives, experiences, and ideas to your organization, which, in turn, promotes innovation and creativity.
3. Global business needs: If your company has international operations or clients, sourcing global candidates can help you better understand and serve these markets.
4. Talent acquisition strategy: If your company has a global talent acquisition strategy, sourcing global candidates can be an integral part of it.

How to Source Global Candidates

1. Define your requirements: Clearly define the skills, qualifications, and experience required for the role, as well as any specific requirements for international candidates.
2. Choose the right platforms: Utilize global job boards, professional networks, and social media platforms to reach a wider audience.
3. Leverage global recruitment agencies: Partner with global recruitment agencies or staffing firms that have expertise in sourcing international candidates.
4. Utilize employee referrals: Encourage your current employees to refer candidates from their global networks.
5. Consider language requirements: Ensure that language requirements are clearly defined and that candidates have the necessary language skills.
6. Research visa requirements: Understand the visa requirements for international candidates and ensure that they can work in the US.
7. Develop a global employer brand: Establish a strong employer brand that appeals to international candidates and showcases your company's culture and values.

Best Ways for Sourcing Global Candidates

1. Cultural sensitivity: Be aware of cultural differences and adapt your recruitment strategy accordingly.
2. Language support: Provide language support for candidates who may not speak the dominant language of the region.
3. Global recruitment tools: Utilize global recruitment tools, such as applicant tracking systems (ATS) and candidate management systems, to streamline the recruitment process.
4. Compliance with regulations: Ensure compliance with all relevant regulations, including visa requirements, employment laws, and tax laws.
5. Onboarding and Support: Provide comprehensive onboarding and support to global candidates, ensuring their successful integration into the company.

Challenges and Solutions

1. Language barriers: Provide language support or utilize translation services to overcome language barriers.
2. Cultural differences: Be aware of cultural differences and adapt your recruitment strategy accordingly.
3. Visa requirements: Research visa requirements and ensure that candidates have the necessary documentation to meet them.
4. Time zone differences: Utilize technology, such as video calls or conference calls, to overcome time zone differences and facilitate effective communication.

By following these guidelines and best practices, you can successfully source global candidates for U.S.-based roles and build a diverse and talented team.

US Work Authorization Types.

The following is a concise overview of common US work authorization types:

1. H1B Visa

- Specialty occupation: For foreign workers in specialty occupations requiring a bachelor's degree or higher.
- Sponsorship: Requires employer sponsorship.
- Duration: Typically valid for 3 years, extendable to 6 years.

2. Optional Practical Training (OPT)

- Students and recent graduates: For international students and recent graduates to work in their field of study.
- Duration: Typically valid for 12-36 months.

3. Employment Authorization Document (EAD)

- Work permit: Allows certain non-citizens to work in the US for a specific period.
- Eligibility: Varies depending on immigration status, such as asylum seekers or spouses of certain visa holders.

4. Green Card (Permanent Resident Card)

- Permanent residence: Grants permanent residence in the US.

- Pathways: This can be obtained through family sponsorship, employment, or other categories.
- Rights and benefits: Similar to US citizens, with some exceptions.

5. US Citizens

- No restrictions: US citizens have no restrictions on their ability to work in the US.
- Preferred for clearance roles: US citizenship is often required or preferred for roles needing security clearance.

These work authorization types have distinct requirements, benefits, and limitations. Understanding these basics can help navigate the complexities of US immigration law.

RISKS AND OPPORTUNITIES WITH VISA-DEPENDENT CANDIDATES.

When hiring visa-dependent candidates, consider the following:

RISKS.

1. Visa uncertainty: Delays or denials can impact candidate availability and job start dates.
2. Sponsorship costs: Employers may incur costs associated with sponsoring visas.
3. Retention challenges: Visa-dependent candidates may face uncertainty about their future in the United States, which could impact job satisfaction and retention.
4. Compliance complexities: Employers must comply with complex and time-consuming immigration regulations.

OPPORTUNITIES.

1. Access to global talent: Hiring visa-dependent visa-dependent candidates can provide access to a broader pool of skilled and qualified professionals from different regions and parts of the world, leading to increased diversity.
2. Diversity and inclusion: Visa-dependent candidates can bring diverse perspectives, experiences, and cultural backgrounds to your organization. When candidates from other regions are involved, they

feel included and valued and are more likely to give their best, knowing that you are not biased.

3. Global competitiveness: Attracting international talent can enhance your company's global competitiveness and innovation. When a company has a diverse workforce with varied backgrounds, experiences, and opinions, it tends to grow even faster, which makes it better than its rivals. Soon, other companies will also reach their standards and not compete with them.

4. Long-term potential: Visa-dependent candidates may eventually become permanent residents or citizens, providing long-term value to your organization.

Mitigating Risks and Maximizing Opportunities

1. Develop a clear immigration strategy: Establish a well-defined process for sponsoring and managing visa-dependent candidates.

2. Communicate openly: Keep candidates informed about the visa process and any potential challenges that may arise.

3. Plan for contingencies: Develop contingency plans to address potential visa delays or denials.

4. Foster an inclusive culture: Create a welcoming and inclusive work environment that supports visa-dependent candidates and promotes diversity and inclusion.

By understanding the risks and opportunities associated with visa-dependent candidates, you can make informed decisions and develop strategies to attract, retain, and support international talent.

Evaluating Immigration Eligibility During Screening

To evaluate immigration eligibility during the screening process, consider the following steps:

Determine Eligibility Requirements

1. Understand sponsorship needs: Determine whether the role requires sponsorship for a work visa or if the candidate already has existing work authorization.

2. Know the company's immigration policies: Familiarize yourself with your company's immigration policies and procedures. Do not act unthinkingly.

Ask Relevant Questions

1. Current work authorization: Ask candidates about their current work authorization status in the US.
2. Visa status: Inquire about their visa status, if applicable.

Assess Eligibility

1. Review immigration documents: Request documentation, such as a valid visa or work permit, to verify the candidate's immigration status.
2. Evaluate visa type: Determine if the candidate's visa type allows them to work in the US in the specific role.
3. Consider sponsorship feasibility: Assess the feasibility of sponsoring the candidate for a work visa, if necessary.

Consult with Immigration Experts

1. Internal immigration team: If available, consult with your company's internal immigration team.
2. External immigration counsel: Consider consulting with external immigration counsel for complex cases or guidance.

Practices to achieve this include:

1. Consistency: Apply the same evaluation criteria to all candidates.
2. Fairness: Ensure that immigration eligibility evaluations are fair and unbiased.
3. Compliance: Verify that your evaluation process complies with relevant laws and regulations.

By following these steps and practices, you can effectively evaluate immigration eligibility during the screening process and ensure compliance with immigration regulations.

USING ETHICAL AND INFORMED PRACTICES WHEN SOURCING INTERNATIONALLY.

When sourcing candidates internationally, it is essential to adopt ethical and informed practices to ensure fairness, transparency, and compliance with local laws and regulations. Here are some guidelines to follow:

Understand Local Laws and Regulations

1. Research local employment laws: Familiarize yourself with employment laws, regulations, and cultural norms in the countries you're sourcing from.
2. Ensure compliance: Verify that your sourcing practices comply with local laws and regulations.

Be Transparent About Job Opportunities

1. Clearly describe job requirements: Accurately represent job responsibilities, expectations, and requirements.
2. Disclose compensation and benefits: Provide clear information about compensation, benefits, and any other relevant employment details.

Respect Cultural Differences

1. Cultural sensitivity: Be aware of cultural differences and adapt your sourcing approach accordingly.
2. Language support: Provide language support for candidates who may not speak the dominant language of the region. Get them an interpreter if possible.

Prioritize Candidate Experience

1. Communicate effectively: Keep candidates informed about the hiring process and provide them with regular updates.
2. Respect candidate time: Be mindful of time zone differences and schedule interviews at convenient times for both parties. The fact that you want to hire them does not give you the right to treat them anyhow or call them at any time of your choice.

Ensure Fairness and Equity

1. Standardized evaluation criteria: Use standardized evaluation criteria to ensure fairness and equity in the hiring process.
2. Avoid bias: Be aware of potential biases and take steps to mitigate them. Do not employ based on skin color, hair color, language, accents, etc.

Consider Partnering with Local Experts

1. Local recruitment agencies: Partner with local recruitment agencies or staffing firms that have expertise in the region.
2. Immigration counsel: Consult with immigration counsel to ensure compliance with local immigration laws and regulations.

By following these guidelines, you can ensure that your international sourcing practices are ethical, informed, and respectful of local laws, regulations, and cultural norms.

WHY IMMIGRATION KNOWLEDGE MATTERS IN SOURCING FOR US ROLES.

Immigration knowledge is crucial when sourcing candidates for US roles for the following reasons:

Ensuring Eligibility

1. Verify work authorization: Confirm candidates have the necessary work authorization to be employed in the US.
2. Understand visa types: Familiarize yourself with various visa types, such as H-1 B, OPT, or Green Card, and their specific requirements.

Avoiding Hiring Challenges

1. Prevent ineligible candidates: Identify candidates who may not be eligible for employment in the US early in the process.
2. Streamline the hiring process: Knowing immigration requirements helps you focus on candidates who are more likely to be hireable.

Compliance with Regulations

1. Adhere to immigration laws: Ensure your hiring practices comply with US immigration laws and regulations.
2. Mitigate risks: Understand potential risks associated with hiring international candidates and take steps to mitigate them.

Effective Talent Acquisition

1. Targeted sourcing: Use immigration knowledge to target candidates who are more likely to be eligible for employment in the US.

2. Improved candidate experience: Provide candidates with accurate information about immigration requirements and processes.

By possessing immigration knowledge, you can efficiently source and hire top talent from a global pool while ensuring compliance with US immigration regulations.

GLOBAL SOURCING: When It Works and When It Doesn't.

Global sourcing can be an effective strategy for identifying top talent, but its success depends on several key factors. It is the practice of searching for and recruiting talent from a worldwide pool of candidates, most often to fill specialized or "hard to find" skill sets while navigating the different immigration laws, regulations, and cultural differences.

This is when global sourcing works well and when it may not:

When Global Sourcing Works Well:

1. Remote roles: US clients open to remote roles can benefit from global sourcing and accessing a broader talent pool.
2. Niche skill sets: Positions requiring specialized or hard-to-find skills can be filled through global sourcing, even if candidates are not locally based.
3. Contractors for short-term roles: Global sourcing can be effective for short-term, project-based, project-based roles where time zone flexibility is beneficial.

When Global Sourcing Doesn't Work Well

1. Strict clearance requirements: Clients with strict clearance or security requirements may require candidates with specific citizenship or residency status.
2. Government contracts: State or federal projects often have strict requirements for US citizenship or specific clearance levels, limiting the pool of eligible candidates.
3. Companies without visa compliance infrastructure: Organizations without the necessary infrastructure or expertise to manage visa compliance may struggle with global sourcing.

Key Considerations

1. Client Requirements: Understand the client's requirements and constraints before pursuing global sourcing.
2. Visa compliance: Ensure your organization has the necessary infrastructure and expertise to manage visa compliance.
3. Talent pool: Assess the availability of talent globally and the feasibility of finding suitable candidates.

By understanding when global sourcing works well and when it doesn't, you can tailor your recruitment strategy to meet client needs and find the best talent.

DIAGRAM.

WORK AUTHORIZATION FLOWCHART.

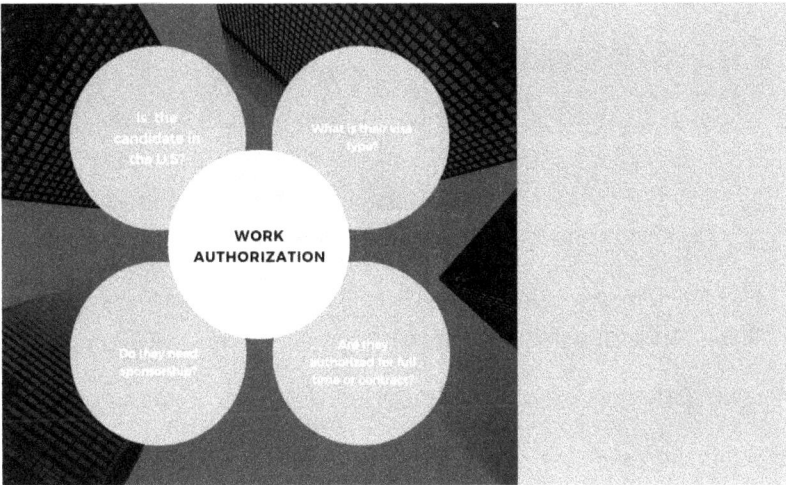

SCREENING TIPS FOR VISA-DEPENDENT CANDIDATES.

When screening visa-dependent candidates, consider the following tips:

Determine Work Authorization

1. Ask about work authorization, for example: "Are you legally authorized to work in the U.S. without sponsorship now or in the future?"

2. Follow-up questions: If not, ask about their current visa type and expiry date.

Verify Visa Details

1. OPT/EAD validity: Verify the validity period and work location flexibility for candidates on OPT or EAD.
2. Visa type and restrictions: Understand any restrictions or limitations associated with their visa type.

Check Client Requirements

1. Review job posting: Check the client's job posting to determine if sponsorship is allowed or required.
2. Ensure alignment: Ensure the candidate's work authorization status aligns with the client's requirements.

Evaluate Candidate Eligibility

1. Assess visa sponsorship needs: Determine whether the candidate requires sponsorship and whether the client is willing to sponsor them.
2. Consider potential risks: Evaluate potential risks associated with the candidate's visa status and plan accordingly.

By following these screening tips, you can effectively evaluate visa-dependent candidates and ensure they meet the client's requirements.

ETHICAL SOURCING CONSIDERATIONS.

When sourcing candidates, particularly those with complex visa situations, it is essential to prioritize ethical considerations. Here are some key principles to follow:

Accurate Representation

1. Truthful about visa status: Never misrepresent a candidate's visa status or work authorization.
2. Clear communication: Ensure clear and accurate communication about visa requirements and sponsorship.

Candidate Privacy

1. Protect sensitive information: Ensure candidate privacy by not sharing visa documents or other sensitive information unless necessary.
2. Confidentiality: Maintain confidentiality and handle candidate information with care.

Sponsorship Promises

1. Client agreement: Don't make promises about sponsorship unless the client has agreed to sponsor the candidate.
2. Transparency: Be transparent about sponsorship requirements and processes.

Regional and Legal Compliance

1. Respect local laws: Respect regional and local laws, regulations, and compliance requirements.
2. Global awareness: Stay informed about global immigration laws and regulations to ensure compliance.

Fair Treatment

1. Equal opportunity: Treat all candidates fairly and without bias, regardless of their visa status.
2. Respect for candidates: Show respect for candidates' time, effort, and circumstances.

Case Example

Finding a Skilled Candidate

A recruiter searched globally and found a skilled candidate based in Africa; the candidate had a valid U.S. Master's degree, making them a strong fit for the role. The candidate was awaiting approval for Optional Practical Training (OPT), which would enable them to work in the US; the recruiter understood the importance of OPT approval and its impact on the candidate's eligibility.

The recruiter clarified timelines with the candidate, including the expected OPT approval date and start date, ensuring everyone was on the same page and helping to manage expectations. The recruiter provided the client with supporting documents, including the candidate's degree and OPT approval documentation, which helped build trust and demonstrate the candidate's qualifications.

The recruiter collaborated with the candidate and client to agree on a start date that worked for everyone, ensuring a smooth transition and minimizing potential delays.

The client appreciated the recruiter's transparency and proactive approach. As a result, the client made an offer to the candidate in advance, securing them for the role early.

NOTE: This example is only scenario-based for training and learning purposes to help us understand the scope better; the real scope can be much deeper and more precise.

Self Assessment

1. What are two visa types that allow international candidates to work in the US?
2. Why might a client reject a candidate based on visa type?
3. What question should you always ask when verifying work eligibility?
4. How can misrepresenting a visa impact your credibility?
5. Why is it essential to clarify sponsorship expectations with your client?

Reflective Prompt: Think of a candidate you've worked with who required sponsorship or had a complex visa situation. What did you learn from that process, and how would you handle it better or differently now?

CHAPTER TEN

TIME ZONE MASTERY AND US GEOGRAPHY FOR RECRUITERS

Objectives of this chapter;

- Understand the four major time zones across the US and how they impact communication.
- Learn how Daylight Saving Time (DST) affects scheduling and availability.
- Identify US tech hubs and the industries they are known for.
- Utilize tools to schedule meetings and interviews effectively.
- Build awareness of regional salary variations and time-based outreach strategies.

THE FOUR MAJOR TIME ZONES IN THE US

There are four primary time zones across the US, they are:

1. Eastern Time (ET): New York, Washington D.C, Florida.
2. Central Time (CT): Texas, Illinois, Minnesota.
3. Mountain Time (MT): Colorado, Arizona, Utah
4. Pacific Time (PT): California, Washington, Oregon.

UNDERSTANDING THE FOUR MAJOR TIME ZONES IN THE UNITED STATES.

The United States spans a vast geographical area, covering six time zones. However, four major time zones are widely recognized: Pacific, Mountain, Central, and Eastern. Now, we will proceed directly to discuss these time zones, exploring their characteristics, notable cities, and differences.

1. **Pacific Time Zone:** The Pacific Time Zone is located on the West Coast of the United States, encompassing states such as California, Washington, Oregon, and Nevada. Some notable cities in this time zone are:

- Los Angeles, California.
- San Francisco, California.
- Seattle, Washington.

- Portland, Oregon.

During standard time, Pacific Time is UTC-8, and during daylight saving time (DST), it observes Pacific Daylight Time, which is UTC-7.

2. **Mountain Time Zone:** The Mountain Time Zone encompasses states such as Arizona, Colorado, Utah, and parts of New Mexico, Oregon, and Texas. Some notable cities in this time zone include:

- Phoenix, Arizona.
- Denver, Colorado.
- Salt Lake City, Utah.
- Albuquerque, New Mexico.

During standard time, Mountain Time is UTC-7; during Daylight Saving Time, it observes Mountain Daylight Time, which is UTC-6. Arizona, except for the Navajo Nation, does not observe DST.

3. **Central Time Zone:** The Central Time Zone spans states such as Texas, Illinois, and Michigan, among others. Some notable cities in this time zone include:

- Houston, Texas.
- Chicago, Illinois.
- New Orleans, Louisiana.
- Minneapolis, Minnesota.

During standard time, Central Time is UTC-6, and during Daylight saving time, it observes Central Daylight Time, which is UTC-5.

4. **Eastern Time Zone:** The Eastern Time Zone covers states such as New York, Florida, and Pennsylvania, as well as parts of several other states. Some notable cities in this time zone include:

- New York City, New York.
- Miami, Florida.
- Boston, Massachusetts.
- Washington D.C.

During standard time, Eastern Time is UTC-5, and during Daylight Saving Time, it observes Eastern Daylight Time, which is UTC-4.

KEY DIFFERENCES AND CONSIDERATIONS.

When working with different time zones, it is essential to consider the time differences and potential impact on communication, scheduling, and productivity. For example;

- When it is 9:00 AM in New York (Eastern Time), it is 8:00 AM in Chicago (Central Time), 7:00 AM in Denver (Mountain Time), and 6:00 AM in Los Angeles (Pacific Time).
- Time zone differences can significantly impact business meetings, conference calls, and deadlines, necessitating careful planning and coordination to ensure effective communication and timely completion of tasks.

What is UTC? UTC stands for Coordinated Universal Time. It is the primary time standard used in modern times to regulate clocks and time globally. UTC is a continuation of Greenwich Mean Time (GMT), serving as a reference point for all time zones.

KEY CHARACTERISTICS OF UTC.

1. UTC is used as a universal standard for timekeeping, ensuring consistent timekeeping across different regions.
2. Time zone reference: It serves as a reference point for all time zones, with time zones expressed as offsets from Coordinated Universal Time (UTC) (e.g., UTC-5 or UTC-2).
3. No daylight saving: UTC does not observe daylight saving time, providing a consistent time standard.
4. Used in technology: It is widely utilized in various technologies, including computer systems, networks, and international communication systems.

IMPORTANCE OF UTC.

1. Global coordination: UTC enables global coordination and synchronization, facilitating international communication, trade, and travel.
2. Consistency: UTC provides a consistent time standard, reducing confusion and errors caused by time zone differences.

3. Scientific applications: It is also utilized in scientific fields, such as astronomy and physics, where precise timekeeping is essential.

For more understanding, let us explain a little further. Pacific Standard Time is referred to as UTC-8 because it is 8 hours behind Coordinated Universal Time. This means that when it is noon UTC, it is 4:00 AM PT. The 'UTC-8' notation indicates the offset from UTC, with negative values representing time zones west of UTC and positive values representing time zones east of UTC.

Time zone awareness in business is a fundamental concept. When working with teams, clients, or individuals across different time zones, it is essential to be mindful of their local time zones and business hours. This includes scheduling interviews, i.e., considering candidates' interviewers' time zones when scheduling interviews. Intake calls, i.e., be aware of the client's or customer's time zone when scheduling intake calls or meetings. Debrief sessions, i.e., scheduling debrief sessions at a time that works for all parties involved, taking into account their respective time zones.

By being aware of the time zones and adapting to local business hours, you can ensure that all parties are on the same page and can participate in meetings and calls without any inconvenience, thereby improving communication, avoiding delays and misunderstandings caused by time zone differences, thereby increasing productivity, foster a more collaborative and inclusive work environment by respecting the time zones and work schedules of all team members.

IMPACT OF TIME ZONES ON COMMUNICATION.

Time zones can significantly impact communication, particularly in global or remote work settings. Some impacts are:

1. Scheduling challenges: Different time zones can make scheduling meetings, calls, or video conferences quite challenging, as it requires careful planning to accommodate all parties.
2. Time differences: Time zone differences can also lead to delays in responses, decision-making, or problem-solving, potentially affecting productivity and collaboration.

3. Coordination difficulties: Time zones can create coordination challenges, especially when working with teams or partners across multiple time zones.
4. Miscommunication risks: It can also increase the risk of miscommunication, such as scheduling conflicts or missed deadlines.

To help reduce these challenges, you may want to try the following;

- Using time zone-friendly tools: Make use of tools like world clocks, time zone converters, or scheduling software to facilitate coordination.
- Setting clear expectations: Establish clear expectations for response times, meeting schedules, and deadlines to avoid misunderstandings.

Being flexible: Be flexible and accommodating when scheduling meetings or calls, considering the time zones of everyone involved. Adapting means being able to accommodate different schedules, finding mutually convenient times, being open to non-traditional work hours, and not always working from 9:00 AM to 5:00 PM.

HOW DAYLIGHT SAVING TIME (DST) AFFECTS SCHEDULING AND AVAILABILITY.

DST can affect scheduling and availability in several ways, some of which include:

1. Time changes: Daylight Saving Time (DST) begins and ends on specific dates, causing clocks to shift forward or backward, which can disrupt schedules.
2. Temporary time zone shifts: During daylight saving time (DST), some time zones may experience a temporary one-hour shift, affecting scheduling and communication.
3. Confusion and errors: DST changes can lead to confusion and errors if not accounted for in scheduling and communication.

To mitigate these effects:

1. Account for Daylight Saving Time (DST) in scheduling; consider DST changes when scheduling meetings, calls, or deadlines to ensure accuracy.

2. Use DST-aware tools; utilize tools that automatically adjust for Daylight Saving Time (DST) changes, such as digital calendars or scheduling software.
3. Communicate clearly; communicate DST changes and their impact on schedules to avoid confusion.

Understanding DST;

Daylight Saving Time (DST) is the practice of temporarily advancing clocks by one hour during the summer months, allowing people to make the most of the sunlight during their waking hours. It typically starts on the second Sunday in March and ends on the first Sunday in November. During this time, clocks are moved an hour ahead in the spring (spring forward) and an hour back in the fall (fall back). Not all states in the United States observe Daylight Saving Time (DST). For example, most of Arizona (except for the Navajo Nation) does not observe Daylight Saving Time (DST).

US TECH HUBS.

The United States is home to a diverse range of tech hubs, each with its unique strengths, industries, and specializations. Here, we will discuss some of the most prominent tech hubs, exploring their industries, notable companies, and what they are known for.

1. SILICON VALLEY, CALIFORNIA.

Silicon Valley is arguably the most iconic tech hub in the world. It is located in the San Francisco Bay Area, a hotbed for startups, venture capitalists, and tech giants. Some notable companies include:

- Apple: Known for its innovative consumer electronics and software.
- Google: A leader in search, advertising, and emerging technologies like AI and quantum computing.
- Facebook: A social media platform with a strong focus on online communities and digital advertising.
- Tesla: A pioneer in electric vehicles and clean energy solutions.

Silicon Valley is known for its:

- Software Development: From operating systems to apps, Silicon Valley is home to some of the world's most influential software companies.

- Artificial Intelligence: The region is also a hub for AI research, development, and application.
- Hardware Innovation: Silicon Valley is home to many outstanding hardware companies, including semiconductor manufacturers and device makers.

2. SEATTLE, WASHINGTON:

Seattle has also emerged as a significant tech hub, driven by the success of companies like:

- Amazon: The e-commerce app/platform is also a leader in cloud computing, AI, and voice assistants.
- Microsoft: A software giant with a strong focus on enterprise solutions, gaming, and AI.

Seattle is known for its:

- E-commerce: Amazon's presence has made Seattle a hub for e-commerce innovation.
- Cloud computing: This region is home to many cloud computing companies, including Amazon Web Services (AWS).
- Gaming: Seattle's gaming industry is thriving, with companies like Microsoft's Xbox and gaming studios.

3. BOSTON, MASSACHUSETTS.

Its prestigious universities and research institutions drive Boston's tech scene. Some notable companies based in Boston include:

- Akamai: A leader in content delivery networks and cybersecurity.
- HubSpot: A marketing and sales software company known for its inbound marketing approach.

Boston is known for its:

- Biotechnology: The region is home to many biotech companies, research institutions, and hospitals.
- Fintech: Boston's financial sector and tech industry intersect in the fintech space.

- Cybersecurity: The region has a strong presence of cybersecurity companies and research institutions.

4. NEW YORK CITY, NEW YORK.

A diverse economy, a robust talent pool, and an entrepreneurial spirit characterize New York City's tech scene. Some notable companies based in New York City include:

- Google: The search engine has a significant presence in New York City, with a focus on advertising and media.
- Facebook: The social media company has a large office in New York City, focusing on advertising and developing products.

New York City is known for its:

- Fintech: The city's financial sector and tech industry intersect in the fintech space.
- Media and entertainment: New York City is a hub for media, entertainment, and digital content creation.
- E-commerce: The city's diverse economy and consumer market make it an attractive location for e-commerce companies.

5. AUSTIN, TEXAS.

Austin's tech scene is driven by its vibrant startup ecosystem, affordable cost of living, and a strong presence of major tech companies. Some notable companies based in Austin include:

- Dell: A leading PC manufacturer with a strong presence in Austin.
- Google: Apart from NYC, this prestigious search company also has a branch in Austin, which focuses on hardware and software development.

Austin is known for its:

- Software development: The city is home to many software companies, from startups to enterprise players.
- Semiconductors: Austin's semiconductor industry is thriving, with companies like Samsung and NXP Semiconductors.
- Gaming: The city boasts a thriving gaming industry featuring prominent companies like Blizzard Entertainment and Electronic Arts.

6. SAN FRANCISCO, CALIFORNIA.

San Francisco's tech scene is driven by its proximity to Silicon Valley (see no. 1), diverse talent pools, and the love of entrepreneurship. Some notable companies based in San Francisco include:

- Twitter (X): A social media platform with a strong focus on real-time communication.
- Airbnb: A pioneer in the sharing economy connecting travelers with unique accommodations.

San Francisco is known for its:

- Software development: The city is home to many software companies, from startups to enterprise players.
- Fintech: San Francisco's financial sector and tech industry, just like in previous regions, also intersect in the fintech space.
- Travel and hospitality: The city's tourism industry and tech companies like Airbnb have created a thriving travel and hospitality tech sector.

7. LOS ANGELES, CALIFORNIA.

Los Angeles' tech scene is driven by its creative industries, entertainment sector, and growing startup ecosystem. Some notable companies based in LA include:

- Snapchat: An app/ social media platform used to text, send streaks, watch videos, etc.

Los Angeles is known for:

- It's creative industries.
- Entertainment.
- Social media.

TOOLS FOR SCHEDULING MEETINGS AND INTERVIEWS.

When scheduling interviews or meetings, it is sometimes essential to have tools to assist with these tasks, as relying solely on your memory can lead to errors and forgetfulness. Some of the tools you would need are:

1. Calendly: It helps automate scheduling and eliminate email back-and-forth. It also integrates with your calendar to avoid conflicts.
2. Doodle: It helps to create polls to find the best time for meetings. It also efficiently schedules meetings with multiple participants.
3. Google Meet/Schedule: It helps to schedule video meetings directly from Google Calendar. It also integrates with Google's suite of productivity tools.
4. Microsoft Outlook: It helps to schedule meetings and manage calendars. It also integrates with Microsoft Teams for video conferencing.
5. ScheduleOnce: It automates scheduling and reduces no-shows. It integrates with various calendar platforms.
6. When2Meet: This tool creates grids to help you find the best time for meetings. It is straightforward and easy to use.
7. Acuity scheduling: It automates appointment scheduling and integrates with various payment gateways.
8. Setmore: It helps with scheduling appointments and meetings and also customizes booking pages and forms.

These tools make work easier for you, speeding up your scheduling processes, saving you time and energy, and enhancing your productivity.

AWARENESS OF REGIONAL SALARY VARIATIONS AND TIME-BASED OUTREACH STRATEGIES.

When working with clients or teams, especially in this context where they are in different regions of the world. It is essential to be aware of:

Regional Salary Differences.

1. Cost of living differences: Salaries can vary significantly depending on the region's cost of living.
2. Industry standards: Different regions may have varying industry standards for salaries.
3. Market rates: Researching market rates for specific roles and locations can help ensure fair compensation or payment.

TIME-BASED OUTREACH STRATEGIES.

1. Time zone differences: Consider the time zones of your clients or team members when scheduling meetings or calls.

2. Local business hours: Be aware of local business hours and holidays to avoid scheduling conflicts.
3. Flexible communications: Be flexible with your communication approach to accommodate the different time zones and their respective schedules.

BENEFITS OF THIS AWARENESS:

- Improved relationships: Showing awareness of regional differences can help build trust and stronger relationships.
- Increased productivity: Of course, if there is an improvement and understanding of the different time zones, clients and team members will be able to work within their schedules, leading to increased productivity within the company.

Better decision-making: Considering regional differences can lead to more informed and effective decision-making.

HOW DO YOU ACHIEVE THIS?

1. Research local norms: Stay up-to-date on local market rates, business hours, and cultural norms.
2. Communicate clearly: Communicate your schedule and expectations to avoid misunderstandings.

Why time zone and regional awareness is crucial;

Time zone and regional awareness are crucial because they pay attention to detail and show professionalism. When you respect others' time and schedules, it demonstrates your professionalism and can prevent unnecessary conflicts, miscommunications, and errors. It can also lead to increased productivity as people work at their convenience, and working at convenient times tends to boost productivity and lead to better relationships. It also helps with global communication and cultural sensitivity. Time zone and location awareness can significantly impact your communication efficiency. Calling a candidate too early or emailing a client too late can signal a lack of professionalism.

DIAGRAM

US TIME ZONES.

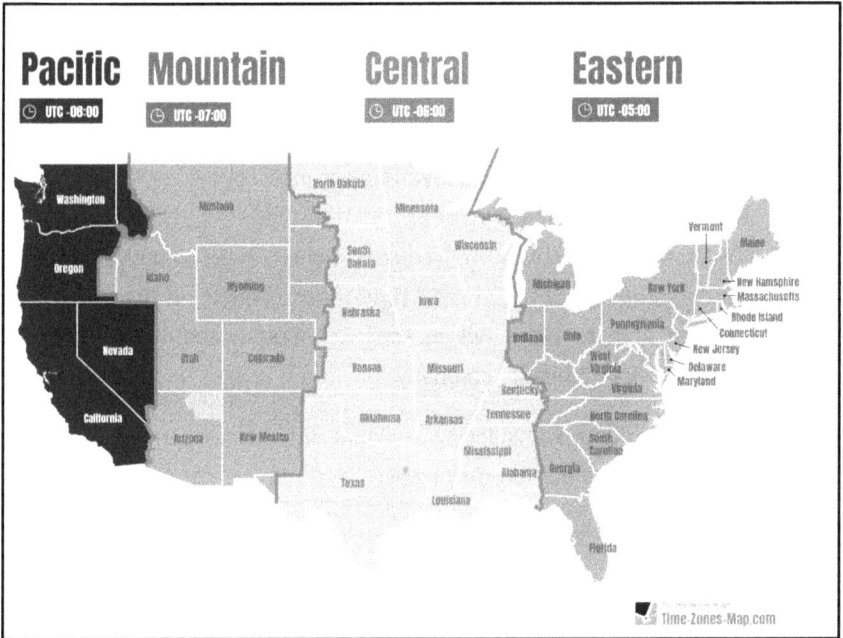

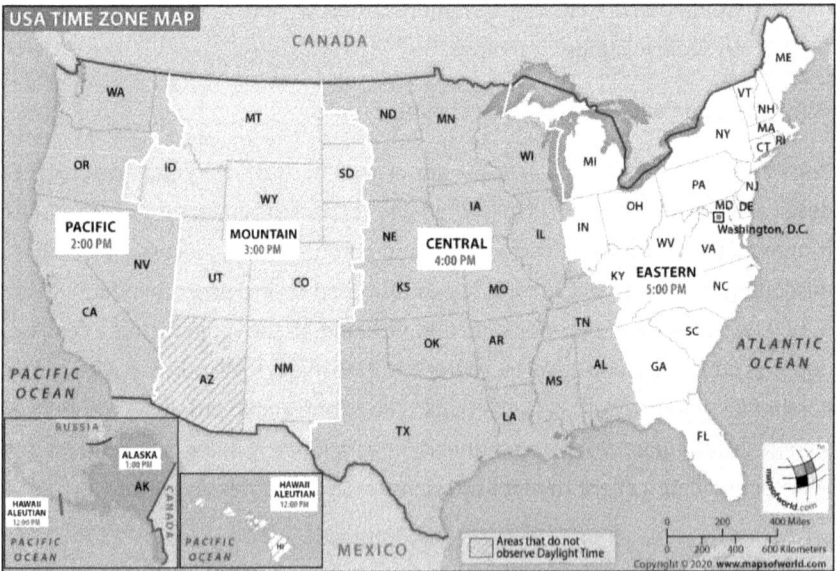

UNDERSTANDING REGIONAL COST OF LIVING AND SALARY DIFFERENCES.

When working with clients or professionals across different regions, it is essential to understand the various costs of living and salary differences. These variations can impact expectations in one way or another and ultimately affect the success of your collaborations.

The cost of living varies significantly across different cities and regions. For instance,

San Francisco and New York are known for their high cost of living, characterized by expensive housing, transportation, and food. As a result, salaries tend to be higher to compensate for these costs.

Midwest cities: Cities in the Midwest, such as Chicago or Indianapolis, have a lower cost of living compared to San Francisco and New York. Salaries in these cities may be lower, reflecting the reduced price of living.

SALARY DIFFERENCES.

Salary expectations can differ significantly depending on the location; these are some factors to consider:

1. Industry Standards: Different industries have varying salary standards, which can vary across regions.
2. Cost of living adjustments: Salaries may be adjusted based on the local cost of living to ensure that professionals can maintain a comparable standard of living.
3. Market rates: Understanding local market rates for specific roles can help you set realistic expectations and negotiate effectively.

So, what impact does this have on client relationships?

Understanding the regional cost of living and salary differences can help you build strong client relationships. These are some reasons:

Setting expectations: By understanding local salary expectations, you can set realistic expectations for your clients and avoid misunderstandings.

- Negotiating smarter: Knowing the local market rates and cost of living can help you negotiate more effectively, ensuring that both parties are satisfied with the agreement.

Building trust: Demonstrating an understanding of regional differences shows that you value your client's needs and are committed to finding mutually beneficial solutions.

CASE EXAMPLE: A recruiter scheduled a screening call for 10:00 AM but failed to confirm the candidate's time zone, resulting in the candidate missing the call. The candidate thought the call was in the Pacific Time Zone (PT), while the recruiter meant the Eastern Time Zone (ET). This led to a slight misunderstanding, and the recruiter had to devise an alternative. He decided to confirm the time zones and also began using calendar invites to avoid misunderstandings.

NOTE: This example is scenario-based training and learning, designed for our purposes to understand the scope better. Thee thee can be much deep-deep deeper

Self-Assessment

1. Name the four major U.S. time zones.
2. What does DST stand for, and when does it typically begin and end?
3. Why is it important to consider geography when discussing salary?
4. Name one tech hub in the US and the industry it is known for.
5. What tool can you use to manage global time differences easily?

Reflective prompt: Think of a time when a scheduling issue created confusion. How could better time zone awareness or tools have helped prevent it?

CHAPTER ELEVEN

TOOLS EVERY RECRUITER SHOULD KNOW

Objectives of this chapter:

- Identify essential productivity and collaboration tools used in global IT recruitment
- Learn how Microsoft Excel and Google Sheets can be used for tracking candidate pipelines
- Understand how PowerPoint and Google Slides support client-facing documentation
- Get introduced to SQL basics for resume database search
- Recognize how documentation and organization affect recruiter performance and trust

ESSENTIAL PRODUCTIVITY AND COLLABORATION TOOLS FOR GLOBAL IT RECRUITMENT.

Here are some key tools to consider:

Recruitment and Talent Acquisition:

- Datapeople: Uses AI to help write better job descriptions and reduce bias.
- Indeed: A globally renowned job aggregator for posting job openings and connecting with top talent.
- Holly: Automates candidate sourcing and shortlisting using AI.
- Toggl Hire: A full cycle recruitment software for assessing candidate's skills and streamlining the hiring process.

Applicant Tracking System (ATS):

- Greenhouse: A comprehensive ATS for tracking candidates, conducting interviews, and managing pipelines.
- TeamTailor: An all-in-one recruitment tool and employer branding ATS.

- Workable: Covers all hiring needs in a single place, suitable for established HR teams.

Video Interviewing and Assessment

Spark Hire: Enables one-way video interviews to assess candidates' soft skills.

- HireVue: Offers text message integrations, skills tests, and integration with HR stacks.
- VidCruiter: An ATS add-on for planning, conducting, and analyzing video interviews.

Social Media Management.

- Zoho Social: Schedules, posts, and shares job openings across social media platforms.
- Hootsuite: A social media scheduling tool with deep analytics and social listening.

Productivity and Collaboration.

- Toggl Plan, Asana, or Trello: Task management tools for staying productive and organized.
- Recruiting CRM software: Streamlines candidate communication and pipeline management.

Global Hiring and Compliance.

- Deel: Simplifies hiring and retaining international workers with compliance and payment management.
- Omnipresent: A UK-based international HR solution for companies.
- Oyster: A global employment solution with compliant, automated hiring and local insights.

THE IMPORTANCE OF TOOLS IN GLOBAL RECRUITMENT.

Tools are essential in everything as they make our work easier. Recruiters can operate more efficiently, document processes, and streamline operations. They will be rested and explained below:

1. Efficiency and Productivity: Tools help recruiters streamline processes, automate repetitive tasks, and manage their time more effectively.

This enables them to focus on high-value activities such as building relationships with clients and candidates.

2. Data-Driven Decision Making: Many recruitment tools provide valuable insights and analytics that can inform strategic decisions. By leveraging these tools, recruiters can make data-driven that enhance their recruitment strategies.

3. Enhanced Candidate Experience: Tools can help recruiters personalize interactions with candidates, ensuring a more engaging and positive experience. This is crucial in attracting and retaining top talent in a competitive market.

4. Client Satisfaction: By utilizing tools to manage client relationships and deliver high-quality candidates, recruiters can significantly improve client satisfaction. This leads to stronger, more lasting relationships and increased business opportunities.

5. Competitive Advantage: In a global market where competition is fierce, having the right tools can provide a significant competitive advantage. Recruiters who are fluent in using these tools can operate more effectively and deliver better results than those who are not.

Now, we will discuss some key tools for recruitment that were not mentioned earlier. Those tools include Microsoft Office, Google Workspace, and Structured Query Language.

Microsoft Office.

The Microsoft Office Suite is a powerful toolset that can help recruiters streamline their workflow, improve productivity, and enhance client communication. The company "Microsoft" has several apps under its name, each of which helps in different ways. Some of these apps are;

- Microsoft Office
- Microsoft teams
- Microsoft one drive
- Microsoft SharePoint
- Microsoft Dynamics
- Microsoft power bi
- Microsoft Azure
- Microsoft Visio

- Microsoft Project

This isn't everything under Microsoft, but we will stop here for now. Now, let's head to the main thing.

Microsoft Office is a productivity app. It includes the following;

- Word (word processing)
- Excel (spreadsheets)
- Powerpoints (presentations)
- Outlook (email and calendar)
- Access (database and management)
- Publisher (desktop publishing)

In this chapter, we will explore how recruiters can utilize Excel, Word, and PowerPoint to optimize their workflow and deliver exceptional results.

Excel for recruitment, Tracking submissions, Scorecards, and status dashboards.

Excel is a powerful tool for recruiters as it enables them to track submissions, build scorecards, and create status dashboards.

Tracking submissions: Create a spreadsheet to track candidate submissions, including candidate's names, job titles, and submission dates. Use filters and sorting to quickly identify top candidates or pending submissions.

Building scorecards: Develop a scorecard template to evaluate candidate qualifications, skills, and experience. Use Excel formulas to calculate scores and rankings, ensuring a fair and objective assessment process.

Status dashboards: Create a dashboard to track the candidate process, including interview schedules, feedback, and offer status. Use conditional formatting to highlight important milestones or deadlines.

Word for recruitment: Formatting Resumes, Cover Letters, and Intake Summaries.

Word is also an essential tool for recruiters, as it enables them to create professional-looking documents that effectively showcase a candidate's qualifications and experience.

Formatting Resumes: Use Word to format your resumes, ensuring consistency and professionalism. Create templates for various job types or industries and utilize styles to apply consistent formatting quickly across all documents.

Cover letters: Develop cover letter templates that can be edited to specific job openings or candidates at any time to reduce the stress of thinking of a letter format. Use merge fields to insert the candidate's name, job title, and other relevant information that may be needed.

Intake summaries: Create intake summaries to capture job requirements, client needs, and candidate profiles. Use Word's outlining feature to organize information and ensure a clear summary.

Powerpoint for recruitment; Presenting Candidate Pipelines or Market Analysis to clients,

PowerPoint is a powerful tool for recruiters, as it helps them present candidates' pipelines, market analysis, and other relevant information to clients.

Candidate pipelines: Create presentations to showcase candidates' candidate pipelines, highlighting candidate qualifications, experience, and fit for the role. Use slides to summarize candidate profiles, skills, and achievements.

Market analysis: Develop presentations to showcase market trends, competitor analysis, and industry insights. Use charts, graphs, and other visual aids to illustrate key points and support recommendations.

Client presentations: Use PowerPoint to provide regular updates to clients on candidate progress, interview feedback, and offer status.

BENEFITS OF USING MICROSOFT OFFICE SUITE.

- Improved productivity; Streamlined workflow and reduced administrative tasks.
- Enhanced client communication: Present information in a clear, concise, and professional manner.
- Data analysis: Use Excel to analyze data, identify trends, and make informed decisions.

Google Workspace.

Google Workspace (formerly G Suite) offers a range of tools that can enhance recruitment processes. The company "Google," just like Microsoft, also has multiple platforms under it, some of which include:

- Google workspace.
- Google Drive
- Google slides
- Google docs
- Google sheets
- Google slides
- Google forms
- Google sites
- Google Calendar
- Google meet
- Google chat.

However, we will be discussing just a few for now.

Google Workspace is made up of:

- Google Docs (word processing)
- Google Sheets (spreadsheets)
- Google slides (presentations)
- Google Drive (cloud storage)
- Google Gmail (email)

Google sheets. Spreadsheets.

It is an essential tool for recruiters, and it's used for:

1. Candidate tracking: Create a spreadsheet to track candidate applications, including contact information, resumes, and interview notes.
2. Job posting management: Use Google Sheets to manage job postings, including job descriptions, requirements, and deadlines.
3. Interview scheduling: Create a sheet to schedule interviews, including candidates' names, interview dates, and times.
4. Candidate evaluation: Develop a scoring system to evaluate candidates, including things like skills, experience, and fit for the role.

5. Reporting and analytics: Utilize Google Sheets to track key recruitment metrics, including time to hire, source of hire, and candidate quality.
6. Collaboration: Share Google SheetsGoogle Sheets with your team members to collaborate on evaluation, interview notes, and recruitment planning.
7. Data analysis: Use Google Sheets to analyze data, identify trends, and inform recruitment decisions.

Google Docs: Word processing.

Google Docs is a word-processing application that enables users to create, share, edit, and collaborate on documents online. One feature of Google Docs is its real-time collaboration, which allows multiple users to edit it simultaneously, with changes being reflected in real time. It is also cloud-based based, which means that documents are stored in the cloud, allowing access from anywhere and on any device. It also features a sharing and permission system, allowing users to share documents with others and control who can view or edit them. It also features a revision history, which tracks changes made to a document, allowing users to revert to previous versions.

USES IN RECRUITMENT:

1. Job descriptions: Create and share job descriptions with team members and stakeholders.
2. Candidate profiles: Use Google Docs to create and share candidate profiles, including resumes, cover letters, and interview notes.
3. Recruitment reports: Create reports on key recruitment metrics, including time to hire and source of hire.
4. Collaboration: Use Google Docs to collaborate with team members on recruitment planning, evaluation, and other recruitment-related tasks.
5. Documentation is used to take notes on company activities and is also utilized for meeting documentation

Google Slides: Presentations.

Google Slides is a presentation app that enables recruiters to create, edit, and share presentations quickly and efficiently. Here are some ways it is used:

1. Candidate presentations: Create presentations to showcase the candidate's qualifications, experience, and fit for a role.

2. Client Presentations: Develop presentations to highlight recruitment progress, candidate quality, and other relevant information.
3. Team meetings: Use Google Slides to present updates, discuss recruitment strategies, and collaborate with team members.
4. Conduct interviews and create presentations to introduce candidates to the company, role, or team.

Recruiter Use Case Examples

Here are some practical examples of how recruiters can use some of these tools, ranging from Microsoft to Google, to streamline their workflow and improve productivity:

EXCEL TRACKER.

1. Automatically calculating submittals per week: Create an Excel tracker that calculates the number of submittals per week, helping recruiters monitor their productivity and meet targets.
2. Response rates: Use Excel formulas to calculate response rates, enabling recruiters to identify areas for improvement and those that are developing rapidly.

GOOGLE SHEETS FOR REMOTE COLLABORATION.

1. Real-time updates: Use Google Sheets to share updates with remote team members in real-time, ensuring everyone is on the same page.
2. Collaborative tracking: Collaborate on tracking candidate progress, job openings, and recruitment metrics to promote transparency and teamwork.

POWERPOINT DECK FOR CANDIDATE PITCH.

1. Visual insights: Create a PowerPoint deck to pitch shortlisted candidates, incorporating visual insights such as charts, graphs, and infographics to highlight candidate qualifications and experience.
2. Candidate comparison: Use PowerPoint to compare candidates, showcasing their strengths, weaknesses, and fit for the role.

Benefits of Using These Tools

1. Increased productivity: Automate calculations, track progress, and collaborate in real-time, saving time and effort.
2. Improved decision-making: Use data-driven insights to inform decision-making, such as candidate selection and recruitment strategies.
3. Enhanced presentation: Create visually appealing presentations to pitch candidates, compellingly showcasing their qualifications and experience.

SQL for Recruiters: Getting Started

SQL (Structured Query Language) is a powerful tool that enables recruiters to efficiently manage and analyze data. Here is how basic SQL skills can benefit recruiters:

Benefits of SQL for Recruiters:

1. Querying internal databases: Use SQL to query internal databases for candidates, retrieving relevant information such as contact details, skills, and experience.
2. Filtering candidates: Filter candidates based on specific criteria, such as location, skills, experience, and availability, to ensure a more targeted recruitment approach.
3. Avoiding ATS duplicates: Utilize SQL to efficiently search for candidates, minimizing duplicates and reducing the likelihood of contacting the same candidate multiple times.

Basic SQL Skills for Recruiters

1. SELECT statements: Use SELECT statements to retrieve specific data from databases, such as candidate names, contact information, and skills.
2. WHERE clauses: Use WHERE clauses to filter data based on specific conditions, such as location or experience.
3. AND/OR operators: Use AND/OR operators to combine multiple conditions and refine search results.

How to use SQL.

1. Online resources: Utilize online resources, such as tutorials and courses, to learn basic SQL skills.
2. Practice: Practice writing SQL queries using sample databases or internal databases.
3. Collaboration: Collaborate with IT teams or data analysts to learn from their expertise and gain insights into database management.

Tips for Recruiters when using the SQL.

1. Start small: Begin with simple SQL queries and gradually build complexity as you become more comfortable with the language.
2. Focus on relevance: Focus on retrieving relevant data that can inform recruitment decisions.
3. Stay up to date: Keep up with changes to databases and SQL syntax to ensure continued effectiveness.

By acquiring basic SQL skills, recruiters can efficiently manage and analyze data, streamline recruitment processes, and make more informed decisions.

DIAGRAM. SCREENSHOT SERIES OF SAMPLE SQL QUERY

```
DECLARE @DynamicPivotQuery AS NVARCHAR(MAX)
DECLARE @ColumnName AS NVARCHAR(MAX)
--Get Distinct values of the PIVOT Column
SELECT @ColumnName= ISNULL(@ColumnName + ',','') + QUOTENAME(Course)
FROM (SELECT DISTINCT Course FROM #CourseSales) AS Courses
--Prepare the PIVOT query using the dynamic
SET @DynamicPivotQuery =
  N'SELECT Year, ' + @ColumnName + '
    FROM #CourseSales
    PIVOT(SUM(Earning)
        FOR Course IN (' + @ColumnName + ')) AS PVTTable'
--Execute the Dynamic Pivot Query
EXEC sp_executesql @DynamicPivotQuery
```

100 %   ▾ ◂

Results | Messages

Year	.NET	Java	Sql Server
1	2012	15000.00	20000.00	NULL
2	2013	48000.00	30000.00	15000.00

TOOL FLUENCY.

Tool fluency refers to the ability to effectively and efficiently use various tools, software, and technologies to achieve specific goals or complete tasks.

KEY BENEFITS OF TOOL FLUENCY

Key Benefits of Tool Fluency

Being fluent in various tools can significantly benefit recruiters in several ways. Here are some key advantages:

Faster Response Time to Clients

1. Efficient data management: Quickly access and manage candidate and job data, enabling faster response times to client inquiries.
2. Streamlined communication: Use tools to automate routine communication, such as email templates or chatbots, to respond to clients and candidates promptly, even when you are not with your phones or computers. For example, if you send an email to Flutterwave, you will get a response immediately. That response is most likely from an email template or a chatbot.

Better Organization of Candidate and Job Data

1. Centralized data storage: Store candidate and job data in a centralized location, making it easily accessible and manageable.
2. Data analysis: Use tools to analyze data, identify trends, and gain insights into candidate behavior and job requirements.

Clear, Trackable Communication

1. Automated workflows: Set up automated workflows to ensure consistent and timely communication with clients and candidates.
2. Tracking interactions: Utilize tools to track interactions with clients and candidates, providing a clear audit trail and facilitating more effective relationship management.

Stronger Performance Reporting

1. Data-driven insights: Utilize tools to generate reports and provide data-driven insights into recruitment performance, including time-to-hire and candidate quality.

2. Client reporting: Provide regular reports to clients, showcasing recruitment progress, candidate quality, and other key metrics.

There are other additional Benefits. Some of these include:

1. Increased productivity: Automate routine tasks and focus on high-value activities, such as building relationships with clients and candidates.
2. Improved decision-making: Utilize data-driven insights to inform recruitment decisions, including candidate selection and job offer strategies.
3. Enhanced client satisfaction: Provide excellent client service, responding promptly to inquiries and delivering high-quality candidates.

Case Example: A recruiter in Europe created a candidate scorecard in Excel with auto-color-coded stages and notes for each round. The client was impressed by the transparency and hired the recruiter as a dedicated partner for three more roles.

Reflective Prompt: What tool or platform do you currently underuse in your workflow? What small step could you take this week to explore or improve how you use it?

CHAPTER TWELVE

LEVERAGING AI IN RECRUITMENT

Objectives of this Chapter:

- Understand what AI tools are commonly used in recruitment
- Learn the benefits and risks of using AI for screening and matching
- Recognize how bias and ethical issues can surface in AI-driven processes
- Explore the balance between Automation and human judgment
- Adopt best practices to use AI responsibly in a global recruitment context

Artificial intelligence (AI) is proving its worth to recruitment teams by offering benefits such as increased efficiency, personalization, and data-informed decision-making. 76% of companies predict that their organizations will implement AI technology within the next 12–18 months to stay competitive, according to Gartner.

Not only has AI-augmented the relationship between people and technology, but it has also transformed the role of HR in attracting, engaging, hiring, and retaining talent.

The modern workforce is starting to grasp the power of AI, but conversations about using AI-driven technology in recruitment effectively can vary, even for seasoned HR professionals.

The terms artificial intelligence (AI) and machine learning (ML) are often used interchangeably. Still, there's a key difference between the two concepts: machine learning is a subset of artificial intelligence. Recruiters will find familiarity with the fundamental differences to be helpful in their jobs.

While artificial intelligence enables machines to make decisions and solve complex problems, machine learning (ML) is the process of making machines "intelligent" by feeding them datasets and specific examples. From that information, the machines can learn to detect nuances and patterns on which to base informed decisions. In short, AI is the goal, and ML is one of the ways to achieve it.

Montesa, M. (n.d). Phenom Library. Retrieved from (phenom.com)

AI TOOLS IN RECRUITMENT.

There has been a significant change in the recruitment industry ever since the discovery of Artificial intelligence (AI). AI-powered tools have changed the way recruiters source, screen, and hire candidates, making the process more efficient, accurate, and cost-effective. In this chapter, we will discuss the various AI tools commonly used in recruitment, their benefits, and how they are transforming the recruitment landscape.

AI SOURCING TOOLS

AI sourcing tools are designed to help recruiters identify the most suitable candidates for a specific job opening. These tools utilize machine learning algorithms to analyze large datasets, including social media profiles, resumes, and online information, to identify top talent.

1. SeekOut: SeekOut is a powerful AI sourcing tool that leverages a vast talent pool of over 800 million public profiles. Its AI-powered matching feature helps recruiters find candidates with the right skills, experience, and qualifications.
2. HireEZ: HireEZ is another popular AI sourcing tool that collects candidate profiles from diverse sources on the Open Web. Its AI-powered search feature enables recruiters to quickly and efficiently find qualified candidates.
3. Betterleap: Betterleap is an AI-powered sourcing tool that utilizes natural language search to identify top talent. Its personalized approach to sourcing ensures that recruiters find candidates who align with both the job requirements and the company culture.

AI SCREENING TOOLS.

AI screening tools are designed to help recruiters evaluate candidates more efficiently and accurately. These tools utilize machine learning algorithms to analyze resumes, cover letters, and other application materials, identifying the most qualified candidates.

1. Humanly: Humanly is an AI-powered screening tool that uses conversational AI to handle screening processes. Its advanced analytics

feature provides valuable insights into candidate behavior and performance.

2. HireVue: HireVue is a popular AI screening tool that assists candidates by matching their skills with available job opportunities. Its conversational AI assistant and auto-updation feature on ATS make it a favorite among recruiters.

3. Skima: Skima is an AI-powered screening tool that gets candidate data from ATS. Its targeted search and relevance scoring features enable recruiters to find the most qualified candidates quickly.

AI INTERVIEW TOOLS.

AI interview tools are designed to help recruiters conduct more efficient and effective interviews. These tools use machine learning algorithms to analyze candidate responses, body language, and tone of voice to evaluate their suitability for the job.

1. Jobma: Jobma is a user-friendly AI interview tool that helps prevent unconscious bias and standardize hiring processes. Its live proctoring and AI transcription features ensure that recruiters get a comprehensive view of candidate performance.

2. Metaview: Metaview is an AI-powered interview tool that captures interviews and meetings, generating notes and summaries with AI-powered interview notes.

AI JOB DESCRIPTION WRITING TOOLS.

AI job description writing tools are designed to assist recruiters in creating more effective job postings. These tools utilize machine learning algorithms to analyze job requirements, company culture, and industry trends, generating job descriptions that attract the most suitable candidates.

1. Textio: Textio is a popular AI job description writing tool that helps recruiters create inclusive job posts by selecting the most effective words tailored to the role and location.

2. ChatGPT: ChatGPT is a free AI job description writer that can generate human-like text for job postings. Its ability to understand natural language and generate coherent text makes it a valuable tool for recruiters.

AI is changing the way companies hire people, but it is not completely replacing human recruiters. Instead, AI is helping recruiters do their jobs more efficiently. AI can quickly scan resumes, match skills, and even conduct initial interviews using chatbots. This Automation can save time and effort for recruiters.

However, relying too heavily on AI can cause problems. If the data used to train AI systems is biased, the AI may make unfair decisions, such as discriminating against certain groups of people. AI may also struggle to understand the nuances of human communication, like tone and context, which can lead to misinterpretation.

Human recruiters bring a level of understanding and empathy that AI currently cannot match, and it is unlikely that it will ever be able to match. Humans can assess a candidate's fit with the company culture, evaluate soft skills, and make more informed decisions.

The future of recruitment is likely to involve a combination of AI and human recruiters. AI can handle routine tasks, freeing recruiters to focus on more complex and high-value tasks, such as building relationships with candidates and making strategic decisions.

By balancing AI and human judgment, companies can create a more efficient and effective hiring process that leverages the strengths of both AI and human judgment. This approach can lead to better hiring decisions and a more positive experience for candidates.

AI IN RECRUITMENT: Key Applications

AI is transforming the recruitment process by streamlining tasks, improving efficiency, and providing valuable insights. Some key applications of AI in recruitment include:

1. Resume Screening

AI-powered tools like HireVue, Pymetrics, and Jobscan scan resumes for keywords and qualifications, enabling recruiters to quickly and accurately identify top candidates.

2. Candidate Matching

AI-driven candidate matching uses scoring algorithms to match candidate profiles with job descriptions, ensuring the best fit for the role.

3. Interview Scheduling

AI-powered bots can coordinate availability and calendar bookings, streamlining the interview scheduling process and reducing administrative tasks.

4. Chatbots

Chatbots can respond to basic candidate questions, collect pre-screening information, and provide timely updates, enhancing the candidate experience.

5. Market Insights

AI can predict compensation benchmarks, time-to-fill, and talent availability, providing recruiters with valuable market insights to inform their hiring strategies.

WHERE AI CAN GO WRONG.

AI can make hiring easier, but it can also cause issues. Some of these issues may include:

1. Unfair Bias

If AI learns from biased data, it can mistreat some candidates based on their gender, ethnicity, or age, as that is how it has already been programmed.

2. Missing Good Candidates

AI has been programmed to select candidates based on specific keywords in their resumes, so it might overlook great candidates who do not have the exact keywords in their resumes, even if they are ideally suited for the job.

3. Not Understanding People

AI cannot fully comprehend essential aspects, such as a candidate's personality, motivation, or alignment with the company culture. It is more focused on technical skills or hard skills than soft skills, which are also crucial to the company.

4. Privacy Issues

Using AI in hiring can raise concerns about how candidate data is used and whether candidates are aware of what is happening with their information.

Being aware of these issues can help organizations use AI in a way that's fair and respectful to all candidates.

Bias and Ethics in AI: Understanding the Issues (A general approach).

Artificial Intelligence (AI) is becoming increasingly important in many areas of our lives, including hiring, healthcare, and finance. However, AI systems can also perpetuate biases and raise ethical concerns if it is not designed and used carefully.

What is Bias in AI?

Bias in AI occurs when an AI system makes decisions that unfairly favor or disadvantage certain groups of people. This can happen in various ways, such as:

1. Biased data: If the data used to train or program an AI system is biased, the system will learn and replicate those biases.
2. Flawed algorithms: AI algorithms can be designed in a way that perpetuates biases, even if the data is neutral.

Examples of Bias in AI

1. Hiring bias: An AI-powered hiring tool may be trained on data that reveals a preference for candidates from certain universities or with specific names, potentially leading to unfair treatment of other candidates.
2. Facial recognition bias: Facial recognition systems may be less accurate for individuals with darker skin tones or certain ethnicities, potentially leading to misidentification.

Ethical Issues in AI

1. Lack of transparency: AI systems can be opaque, making it difficult to understand how decisions are made.
2. Accountability: Who is responsible when an AI system makes a mistake or perpetuates bias? No one! So, this is an issue because no one can be held accountable for any problems.
3. Data privacy: AI systems often rely on vast amounts of personal data, raising concerns about data protection and privacy.

Consequences of Bias and Ethical Issues

1. Unfair treatment: Bias in AI can lead to unfair treatment of certain groups, perpetuating existing social inequalities.

2. Loss of trust: If AI systems are perceived as biased or unfair, people may lose confidence in the technology and the organizations using it.
3. Reputational damage: Organizations that deploy biased or unethical AI systems can suffer reputational damage and potential financial losses.

How to Resolve Bias and Ethical Issues

1. Diverse and representative data: Ensure that AI systems are trained on diverse and representative data to minimize bias.
2. Transparency and explainability: Design AI systems that are transparent and explainable, allowing humans to understand how decisions are made.
3. Human oversight: Implement human oversight and review processes to detect and correct bias.
4. Ethical design: Design AI systems with ethics and fairness in mind from the outset.

FINDING THE RIGHT BALANCE BETWEEN AUTOMATION AND HUMAN JUDGMENTS.

Automation and human judgment are both critical in many areas, including hiring, decision-making, and problem-solving. While Automation can bring efficiency and speed, human judgment provides nuance, empathy, and critical thinking.

The Benefits of Automation:

1. Speed and efficiency: Automation can process large amounts of data quickly and accurately, freeing up time for more critical tasks.
2. Consistency: Automation can ensure consistency in decision-making, reducing the risk of human bias and errors.

The Importance of Human Judgment

1. Context and nuance: Humans can understand complex situations and nuances that Automation might miss.
2. Empathy and compassion: Humans can offer empathy and compassion, which are essential in fields such as customer service, healthcare, and social work.
3. Critical thinking: Humans can think critically and make decisions that require creativity, intuition, and experience.

Note that Automation has its advantages, but relying too heavily on Automation has a lot of challenges. For example, it might not fully understand the context of a situation, which would lead to making decisions that are not suitable. It would also be biased and contain numerous errors. It is also very inflexible, which would make it very difficult for it to adapt to changes in circumstances.

The Benefits of Balancing Automation and Human Judgment.

1. Improved decision-making: By combining Automation and human judgment, organizations can make more informed and balanced decisions.
2. Increased efficiency: Automation can handle routine tasks, freeing up humans to focus on more complex and high-value tasks.
3. Enhanced customer experience: By leveraging the strengths of both Automation and human judgment, organizations can provide a better experience for customers and stakeholders.

Achieving the Right Balance

1. Identify tasks that require human judgment: Determine which tasks require empathy, creativity, and critical thinking, and ensure that humans are involved in those areas.
2. Utilize Automation for Routine Tasks: Leverage Automation for routine and repetitive tasks, freeing up humans to focus on more complex tasks.
3. Monitor and adjust: Continuously monitor the balance between Automation and human judgment, making adjustments as needed to ensure that the benefits of both are being realized.

COMPARISON BETWEEN AI AND HUMAN INPUT;

TASK	AI STRENGTH	HUMAN STRENGTH
Keyword matching	Easily achieved	Difficult
Understanding career gaps	Difficult	Easily achieved
Culture fit assessment	Difficult	Easily achieved
Resume volume filtering	Easily achieved	Difficult

HOW TO USE AI RESPONSIBLY IN HIRING.

When using AI in the hiring process, it's essential to do so responsibly to ensure fairness and transparency. Here are some key considerations:

1. Be Open with Candidates

Tell candidates if you are using AI in your hiring process. This builds trust and lets them know what to expect. Transparency is key to maintaining a positive candidate experience.

2. Do not Rely Solely on AI for Decision-Making

Use AI as a tool to help filter candidates, but don't let it make the final decision. AI can help identify top candidates, but human judgment is still necessary to ensure the best fit for both the role and the company.

3. Review Resumes Manually

AI might miss great candidates who do not fit the exact criteria. Review resumes flagged by AI manually to catch potential candidates that AI might have overlooked.

4. Train Your Team

Ensure your team understands how to use AI ethically and complies with local laws and regulations. This includes being aware of potential biases and taking steps to mitigate them.

5. Regularly Evaluate AI Tools

Regularly assess the AI tools you are using to ensure they are fair and unbiased. Check for false positives or negatives and make adjustments as needed to maintain a fair and effective hiring process.

Case Example:

A global staffing firm used an AI tool to match resumes with job openings. However, the tool was set up to favor candidates with specific keywords related to US education. As a result, many qualified candidates from diverse backgrounds were being overlooked.

The company noticed that its shortlists of candidates lacked diversity as everyone there had a US education. This meant that the AI tool was unintentionally filtering out talented individuals who could have been an excellent fit for the job.

The company took steps to fix the issue. They revised the criteria used by the AI tool to be more inclusive and retrained the model. They also added human review checkpoints to ensure that the AI's decisions were fair and accurate.

By making these changes, the company was able to improve diversity in its candidate pool. Clients were also more satisfied with the quality of candidates being presented. This experience taught the company the importance of regularly reviewing and adjusting its AI tools to ensure fairness and effectiveness.

NOTE: This example is scenario-based for training and learning purposes to help us understand the scope better. The real scope can be much deeper and more precise.

Self Assessment

1. Name one benefit and one drawback of using AI in recruitment.
2. Why is it risky to let AI be the sole decision-maker in screening?
3. What's the best practice to ensure fairness when using AI tools?
4. How can bias enter AI recruitment tools?
5. When reviewing AI-screened resumes, what should a recruiter double-check?

CHAPTER THIRTEEN

SCREENING, VETTING AND ETHICS

Objectives of this chapter:

- Understand how to create and follow a structured screening checklist
- Learn how to conduct effective and respectful on-camera interviews
- Know what questions and document requests are appropriate and ethical
- Build trust with clients and candidates through fair vetting practices
- Avoid legal risks and compliance issues across international recruiting

SCREENING.

Screening refers to the process of evaluating and filtering candidates to determine their suitability for a particular role or position. It involves assessing their qualifications, skills, experience, and other relevant factors to identify potential candidates who may be a good fit for the job.

VETTING.

Vetting is a more in-depth process of investigating and evaluating a candidate's background, credentials, and character. It involves verifying the accuracy of the information provided by the candidate and assessing potential risks or concerns that may impact their suitability for the role or position.

While Screening is often used to identify potential candidates, Vetting is used to further evaluate and validate the candidate's credentials and character, often for roles that require high levels of security, trust, or responsibility.

The Importance of Screening and Vetting in Recruitment. With the rise of remote work and global sourcing, recruiters face the challenge of finding the right candidates for their clients while ensuring that they protect both their clients and candidates. This is where Screening and Vetting come in as some crucial steps in the recruitment process that go beyond just ticking boxes.

So, What is Screening and Vetting as a concept of its own?

Screening and Vetting involve a thorough evaluation of a candidate's background, skills, and motivation to ensure that they are a good fit for the role and the company. This process helps recruiters to identify potential risks, verify the candidate's credentials, and assess their suitability for the position.

Why is Screening and Vetting Important?

Screening and Vetting are essential for several reasons:

1. Reducing Risk: A thorough screening and vetting process can help identify potential risks associated with a candidate, such as a history of dishonesty or violent behavior. This can protect the company, its employees, and its clients from potential harm.
2. Verifying Fit: Screening and Vetting enable recruiters to verify that a candidate's skills, experience, and motivation align with the role's requirements and the company's culture.
3. Protecting Clients and Candidates: By conducting thorough screenings and Vetting, recruiters can protect their clients from potential problems, such as hiring someone who is not qualified or who may pose a risk to the company. At the same time, recruiters can also protect candidates from being misrepresented or mistreated.

Challenges of Global Sourcing

With global sourcing, Screening and Vetting become even more challenging to do. Recruiters need to familiarize themselves with various laws, regulations, and cultural norms, which can make it difficult to ensure that the screening and vetting process is both practical and compliant.

Ensuring Ethical Vetting

To ensure that Vetting is done ethically, recruiters need to be aware of the potential problems and take steps to achieve them. This includes:

- Respecting Privacy: Recruiters must ensure that they respect candidates' privacy and refrain from collecting unnecessary information that is unrelated to the job.
- Complying with Laws and Regulations: Recruiters must comply with relevant laws and regulations, including data protection and anti-discrimination laws.

- Being Transparent: Recruiters should be transparent about the vetting process and inform candidates about what information is being collected and how it will be used.

Ways for a Successful Screening and Vetting

To conduct effective and ethical Screening and Vetting, recruiters should follow practices such as:

- Using a combination of methods: Recruiters should use a combination of methods, such as background checks, reference checks, and interviews, to get a comprehensive view of the candidate.
- Verifying information: Recruiters should verify the information provided by the candidate to ensure that it is accurate and truthful.
- Assessing motivation: Recruiters should evaluate the candidate's motivation and suitability for the role and the company culture.

BUILDING A COMPREHENSIVE SCREENING CHECKLIST.

When screening candidates, it is essential to have a clear and comprehensive checklist to ensure you are evaluating the right skills, qualifications, and fit for the role. In this context, we are going to be discussing the breakdown of the key elements to include in your screening checklist:

1. Role Specific Skills.

These are the technical skills required for the job. Examples include:

- Programming languages like Java, Python, or C++
- Cloud platforms like AWS, Azure, or Google Cloud
- Specialized software like Salesforce, HubSpot, or Adobe Creative Cloud

When evaluating role-specific skills, consider:

- Does the candidate have the required technical skills for the job?
- Are their skills up-to-date and relevant to the current industry trends?
- Can they apply their skills to solve real-world problems?

2. Domain Knowledge

In Chapter 2 of this book, we talked about Domain Knowledge, so it is advised you go through the chapter again to understand this better. It refers to the candidate's knowledge and experience in a specific industry or domain. Examples include:

- Healthcare: medical terminology, healthcare regulations, and industry-specific software
- FinTech: financial regulations, payment systems, and fintech-specific technologies
- E-commerce: online marketing, digital payments, and e-commerce platforms

When evaluating domain knowledge, consider the following:

- Does the candidate have relevant experience and knowledge in the industry or domain?
- Can they apply their knowledge to solve industry-specific problems?
- Are they up to date with the latest industry trends and developments?

3. Communication Skills

In Chapter 5 of this book, we also discussed communication skills. It is advised that you also reread it to learn about the different communication skills in Recruitment. Effective communication is critical in any role. When evaluating communication skills, consider the following:

- Clarity: Can the candidate clearly articulate their thoughts and ideas?
- Tone: Is the candidate's tone professional and respectful?
- Professionalism: Does the candidate demonstrate a professional demeanor and attitude?

When assessing communication skills, consider the following:

- How well does the candidate listen and respond to questions?
- Can they effectively communicate complex ideas?
- Are they able to adapt their communication style to different audiences?

4. Work Authorization and Availability

Before moving forward with a candidate, it is essential to confirm their work authorization and availability. Consider:

- Does the candidate have the necessary work permits or visas?

- Are they available to start work on the required date?
- Are they willing to work the required hours or shifts?

5. Compensation Expectations and Flexibility

In Chapter 6 of this book, we discussed compensation models. Understanding a candidate's compensation expectations and flexibility is crucial for ensuring a good fit for both the role and the company. Consider:

- What are the candidate's salary expectations?
- Are they open to negotiation?
- Are they willing to consider other benefits or perks?

6. Notice Period or Current Employment Status

Knowing a candidate's notice period or current employment status can help you plan for the hiring process and potential start date. Consider:

- How much notice does the candidate need to give their current employer?
- Are they currently employed, and if so, what is their expected start date?
- Are there any potential issues with their current employment contract?

7. Cultural Fit and Soft Skills

Cultural fit and soft skills are crucial for a candidate's long-term success in the role and within the company. Consider:

- Does the candidate's values and work style align with the company culture?
- Do they demonstrate soft skills like teamwork, adaptability, and problem-solving?
- Can they work effectively with others and build strong relationships?

When evaluating cultural fit and soft skills, consider the following:

- How well does the candidate fit in with the company culture and values?
- Can they adapt to changing circumstances and priorities?
- Do they demonstrate a positive and professional attitude?

By including these elements in your screening checklist, you can ensure a comprehensive evaluation of each candidate and find the best fit for the role and company.

CONDUCTING EFFECTIVE ON-CAMERA

On-camera interviews have become increasingly popular, especially with the rise of remote work and virtual hiring processes. To ensure a successful on-camera interview, it is essential to follow some guidelines. Here is a detailed breakdown of what and what to consider:

1. Ensure a Suitable Environment

Before starting the interview, ensure the candidate is in a quiet, safe, and well-lit location. This will help minimize distractions and allow the candidate to focus on the conversation.

- Quiet location: Find a room or space with minimal background noise and interruptions.
- Safe location: Ensure the candidate feels comfortable and secure during the interview.
- Well-lit location: Good lighting is essential for video quality. Natural light or soft, indirect lighting works best.

2. Introduction and Role Confirmation

Begin the interview by introducing yourself and confirming the role being discussed. This helps establish a professional tone and ensures the candidate is aware of the position.

- Introduction: Clearly state your name, title, and the company you are representing.
- Role confirmation: Confirm the job title, role, and responsibilities being discussed.

3. Behavioral and Technical Questions

Ask behavioral and technical questions that are directly tied to the job description (JD). This helps assess the candidate's relevant experience, skills, and fit for the role.

- Behavioral questions: Ask questions that explore the candidate's past experiences and behaviors, such as "Tell me about a time when…"
- Technical questions: Ask questions that assess the candidate's technical knowledge and skills, such as "How would you approach…"

4. Observing Soft Skills

Pay attention to the candidate's soft skills, such as:

- Eye contact: Do they maintain good eye contact during the conversation?
- Confidence: Do they demonstrate confidence and self-assurance?
- Professionalism: Do they display a professional demeanor and attitude?

These soft skills are essential for success in most roles and can significantly enhance a candidate's overall fit.

5. Taking Notes and Avoiding Multitasking

Take notes respectfully and avoid Multitasking during the interview. This helps ensure you are fully present and engaged with the candidate.

- Take notes: Jot down important points, answers, and impressions.
- Avoid Multitasking: Refrain from checking emails, messages, or other tasks during the interview.

DIAGRAM.

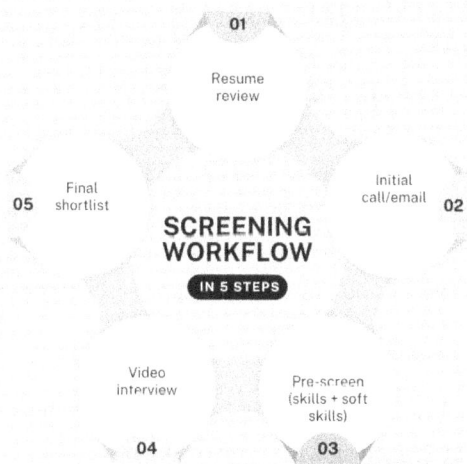

01
Resume review

05 Final shortlist

Initial call/email 02

SCREENING WORKFLOW
IN 5 STEPS

Video interview

Pre-screen (skills + soft skills)

04

03

UNDERSTANDING WHAT IS OKAY TO ASK AND WHAT IS NOT IN INTERVIEWS.

When conducting interviews, it is essential to know which questions are appropriate and which might put you in a difficult situation. Let's break it down:

What is Okay to Ask?

These questions are relevant to the job and help you assess the candidate's qualifications and fit:

1. Work authorization: "Are you legally authorized to work in the U.S.?" This question ensures the candidate has the necessary permissions to work in the country.
2. Notice period: "What is your notice period?" This question helps you understand when the candidate can start working and plan for the hiring process.
3. Relevant experience: "Do you have experience with [X] tool or in [Y] domain?" This question assesses the candidate's relevant skills and knowledge for the role.

What is Not Okay to Ask?

These questions are potentially discriminatory or invasive and should be avoided:

1. Age or marital status: "What is your age/marital status?" This question is not relevant to the job and could lead to age or marital status-based discrimination.
2. Personal documents: "Can you send me your visa, passport, or I-797?" Unless legally required and approved by the client, requesting personal documents can be perceived as invasive or discriminatory.
3. Religion or ethnicity: "What religion or ethnicity are you?" This question is highly sensitive and potentially discriminatory. It is essential to focus on qualifications and skills rather than personal characteristics.

Asking inappropriate questions can lead to the following:

1. Discrimination claims: Candidates may feel that they have been discriminated against, which can result in legal issues and reputational damage.

2. Lack of trust: Candidates may feel uncomfortable or mistrustful of the hiring process, which can impact their experience and perception of the company.
3. Missed opportunities: By focusing on irrelevant questions, you might overlook qualified candidates who could be an excellent fit for the role.

ETHICAL CONSIDERATIONS IN VETTING.

Vetting candidates is a crucial part of the hiring process. However, it is essential to do it ethically and with respect for the candidates. Here are some ethical considerations you should keep in mind:

1. Get Consent Before Recording or Sharing Interviews

When conducting interviews, it is essential to get the candidate's consent before recording or sharing the conversation.

- Recording interviews: If you plan to record the interview, ask the candidate's permission first. This ensures they are aware and comfortable with the recording.
- Sharing interviews: If you plan to share the interview with others, such as colleagues or hiring managers, get the candidate's consent first.

2. Do Not Falsify Resumes, Documents, or Candidate Statements

It is crucial to maintain the integrity of the hiring process by not falsifying any information.

- Accurate representation: Ensure that all information about the candidate is accurate and truthful.
- No fabrications: Do not add or alter information to make a candidate appear more qualified than they are.

3. Never Push Candidates to Lie About Experience

It is essential to respect the candidate's honesty and experience.

- No pressure to exaggerate: Do not push candidates to exaggerate or lie about their experience or qualifications.
- Respect their honesty: If a candidate does not have a particular skill or experience, respect their honesty and don't pressure them to claim otherwise.

4. Respect Rejection; Don't Harass or Guilt Trip

When rejecting candidates, it is essential to maintain a respectful and professional tone in your communication.

- Clear communication: Communicate the rejection clearly and respectfully.
- No harassment: Do not harass or guilt trip candidates into reconsidering or applying for other roles.

5. Share Constructive Feedback

Sharing constructive feedback can be incredibly valuable for candidates, especially those who were rejected.

- Helpful feedback: Provide feedback that is specific, constructive, and actionable.
- Growth opportunity: Help candidates grow and improve by providing feedback that highlights areas for development.

Why Ethical Vetting Matters.

Ethical Vetting is crucial for:

- Building trust: Candidates are more likely to trust your organization if you treat them with respect and professionalism.
- Maintaining reputation: A reputation for ethical Vetting can attract top talent and enhance your organization's reputation.
- Fairness: Ethical Vetting ensures that all candidates are treated fairly and without bias.

Ways for Ethical Vetting

To ensure ethical Vetting, follow these practices:

- Be transparent: Communicate the vetting process and what candidates can expect.
- Respect candidates: Treat candidates with respect and professionalism throughout the vetting process.
- Maintain confidentiality: Keep candidate information confidential and secure.

Case example:

The recruiter asked the candidate to turn on their video during the call, and the candidate was in a car at the time. The hiring manager witnessed this and immediately ended the call. The recruiter did not consider the candidate's environment and safety before asking them to turn on their video. This puts the candidate in an uncomfortable or potentially unsafe situation.

The recruiter learned a valuable lesson from this experience. Now, they:

1. Confirm location and safety: Before every call, the recruiter confirms the candidate's location and ensures they are in a safe and comfortable environment.
2. Set expectations in advance: The recruiter sets clear expectations with the candidate about the call, including whether video will be required and what kind of environment is suitable.

NOTE: This example is scenario-based for training and learning purposes to help us understand the scope better. The real scope can be much deeper and more precise.

Self Assessment.

1. Name three key components of a strong screening checklist.
2. What is one thing you should check before conducting a video interview?
3. Give two examples of questions that are NOT appropriate during Screening.
4. Why is candidate consent necessary before recording or sharing information?
5. How can proper Vetting reduce risk for your client?

Reflective Prompt: Think of a past screening where something went wrong (e.g., poor setup, missed question, wrong fit). What would you improve next time to make your screening process more ethical and practical?

CHAPTER FOURTEEN

CUSTOMIZING STRATEGY FOR EVERY RECRUITMENT

Objectives of this Chapter:

- Learn why each job requirement deserves a customized recruiting strategy
- Identify how different factors affect sourcing, Screening, and candidate management
- Use a strategic intake sheet to capture key hiring criteria
- Understand how to pivot approaches when market conditions or feedback changes
- Build consistency in delivery while staying flexible in execution

WHY EACH JOB REQUIREMENT DESERVES A CUSTOMIZED RECRUITING STRATEGY.

When it comes to recruiting, a one-size-fits-all approach just doesn't cut it. Each job requirement is unique, and the recruiting strategy should be tailored to the specific needs of the role. This is crucial to avoid misunderstandings and to secure the exact candidates you need. Now, let's talk about other reasons why this is so important.

- Unique Job Requirements Demand Unique Approaches

   Every job has its own set of requirements, responsibilities, and challenges. A customized recruiting strategy ensures that you are targeting the right candidates with the right skills, experience, and fit for the role.

- Different roles require different skills: A software engineer role requires different skills and qualifications than a marketing manager role. A customized recruiting strategy takes into account the specific skills and qualifications needed for the job. A sourcing strategy that works for one DevOps position may fail for another with a similar title but a different tech stack or business need.

- Industry-specific knowledge: Certain industries, such as healthcare or finance, require specific knowledge and expertise. A customized recruiting strategy ensures that you are targeting candidates with the proper industry-specific knowledge and experience.

BENEFITS OF CUSTOMIZED RECRUITING STRATEGIES.

A customized recruiting strategy offers several benefits, including:

- Better candidate fit: By matching your recruiting strategy to the specific job requirements, you are more likely to attract candidates who are a good fit for the role. Most of the time, you may not immediately find the exact candidate you need, but it is better than recruiting someone who is not a good fit for the role.
- Improved candidate experience: A customized recruiting strategy ensures that candidates have a positive experience throughout the hiring process, which can enhance your employer's brand and reputation.
- Increased efficiency: By targeting the right candidates with the right skills and experience, you can reduce the time and resources spent on recruiting and hiring. It can also increase job efficiency, as the candidate knows exactly what they are doing and is not working on assumptions.

HOW TO DEVELOP A CUSTOMIZED RECRUITING STRATEGY.

To develop a customized recruiting strategy, consider the following steps:

Define the job requirements: Clearly define the job requirements, including the necessary skills, experience, and qualifications for the role.

- Identify the target audience: Identify the target audience for the role, including the type of candidates you are looking for and where they are likely to be found.
- Choose the proper recruitment channels: Choose the recruitment channels that are most likely to reach your target audience, such as social media, job boards, or employee referrals.

- Develop a compelling job description: Develop a compelling job description that accurately reflects the job requirements and responsibilities and appeals to the target audience.

Examples of Customized Recruiting Strategies:

- Social media recruitment: For a role that requires a strong social media presence, you might use social media platforms to reach potential candidates and promote the job opportunity.
- Industry-specific job boards: For a role that requires industry-specific knowledge and expertise, you might post the job on industry-specific job boards or websites.
- Employee referrals: For a role that requires a strong cultural fit, you might use employee referrals to reach candidates who are already familiar with the company culture and values.

Each job requirement deserves a customized recruiting strategy that takes into account the unique needs and requirements of the role. By aligning your recruiting strategy with the specific job requirements, you can attract the right candidates, enhance the candidate experience, and streamline hiring efficiency.

KEY VARIABLES TO CONSIDER WHEN BUILDING STRATEGY.

When developing a recruitment strategy, several key variables should be taken into consideration. These variables can significantly impact the success of your recruitment efforts. Let's talk about each of these variables and explore how they can influence your strategy.

1. Urgency of Hire

The urgency of the hire is a critical factor in determining your recruitment strategy. You need to consider whether the position needs to be filled immediately or if you have time to build a long-term pipeline.

- Immediate start: If the position needs to be filled immediately, you may need to focus on more traditional recruitment methods, such as job boards or employee referrals. This approach can help you find candidates quickly, but it may not be the most effective way to find the best candidate.

- Long-term pipeline: If you have time to build a long-term pipeline, you can focus on more strategic recruitment methods, such as employer branding, social media, and talent pipeline. This approach can help you attract top talent and build a pool of qualified candidates for future openings.

2. Hiring Manager's Expectations

The hiring manager's expectations can also impact your recruitment strategy. You need to understand their level of involvement and expectations for the recruitment process.

Hands-on involvement: If the hiring manager wants to be heavily involved in the recruitment process, ensure they are available for interviews and can provide timely feedback. This approach ensures that the hiring manager is invested in the process and can provide valuable insights into the candidates.

- Hands-off involvement: If the hiring manager prefers a more hands-off approach, you can take the lead in managing the recruitment process and providing regular updates. This approach can help streamline the process and reduce the burden on the hiring manager.

3. Job Market Conditions

Job market conditions can significantly impact your recruitment strategy. You need to consider the demand for the role and the availability of candidates.

- Role in demand: If the role is in high demand, you may need to be more competitive in your recruitment efforts, such as offering higher salaries or more attractive benefits. This approach can help you attract top talent and stay competitive in the market.
- Niche skills: If the role requires niche skills, you may need to focus on more targeted recruitment methods, such as industry-specific job boards or social media groups. This approach can help you reach candidates with the specific skills and experience you need.

4. Candidate Landscape

The candidate landscape can also impact your recruitment strategy. You need to consider the availability of candidates and their level of interest in the role.

- High supply: If there are many qualified candidates available, you can focus on more competitive recruitment methods, such as assessing candidate skills and experience. This approach can help you identify the best candidates and make informed hiring decisions.
- Limited availability: If there are limited candidates available, you may need to focus on more creative recruitment methods, such as employee referrals or social media outreach. This approach can help you reach candidates who may not be actively job searching but could be a good fit for the role.

5. Client Brand & Budget

The client brand and budget can also impact your recruitment strategy. You need to consider the attractiveness of the employer brand and the budget available for recruitment efforts.

- Attractive employer: If the client has a strong employer brand, you can leverage this to attract top talent. This approach can help you attract candidates who are interested in working for a reputable company.
- Limited incentives: If the client has limited incentives, such as a lower salary or fewer benefits, you may need to focus on other aspects of the job, such as the company culture or opportunities for growth and development. This approach can help you attract candidates who are motivated by factors other than salary.

REQUIREMENT INTAKE STRATEGY SHEET: A Simple yet Powerful Tool.

A Requirement Intake Strategy Sheet is a simple form that helps capture essential information about a job requirement. This form ensures that all stakeholders are aligned and that the recruitment process is well-planned and executed. Each section of the form includes some information, and we are going to discuss it as well as its significance.

1. Job Title and Domain

The job title and domain provide context about the role and industry.

- Job title: Clearly defines the role and responsibilities.
- Domain: Specifies the industry or field the role belongs to, for example, the medical field.

2. Required Skills vs. Preferred Skills

Distinguishing between required and preferred skills helps identify the essential qualifications for the role.

- Required skills: Must haves for the role, such as specific technical skills or certifications.
- Preferred skills: Nice to haves that can enhance the candidate's performance, such as soft skills or additional certifications.

3. Key Screening Questions

Key screening questions help evaluate candidates' qualifications and determine their suitability for the role.

- Screening questions: Specific questions that assess candidates' skills, experience, and knowledge.

4. Deal Breakers

Deal breakers are non-negotiable requirements that can immediately disqualify a candidate.

- Deal breakers: Essential requirements that cannot be compromised on, such as specific certifications or experience.

5. Budget Range & Rate Flexibility

Understanding the budget range and rate flexibility helps manage expectations and facilitates effective negotiation with candidates.

- Budget range: The allocated budget for the role.
- Rate flexibility: The degree of flexibility in the budget to accommodate different rates or salaries.

6. TIme Zone/Location Needs

Time zone and location needs impact the recruitment process and candidate selection.

- Time zone: Specifies the required time zone for the role.
- Location: Defines the physical location requirements, whether remote or on-site.

7. Interview Rounds & Format

The interview process and format can significantly impact the candidate's experience and evaluation.

- Interview rounds: The number of interview rounds and their purpose.
- Format: The format of the interviews, such as video, phone, or in person.

8. Submission Format Preferences

Submission format preferences ensure that candidates provide the required information in the desired format.

- Submission format: The preferred format for candidate submissions, such as resume, cover letter, or portfolio.

Benefits of a Requirement Intake Strategy Sheet.

A Requirement Intake Strategy Sheet offers several benefits, including:

- Clear understanding: Ensures all stakeholders have a clear understanding of the job requirements and expectations.
- Streamlined process: Helps streamline the recruitment process by identifying essential requirements and deal breakers.
- Better candidate fit: Increases the likelihood of finding the best candidate fit by capturing specific skills and qualifications.

Things to put in place when using a Requirement Intake Strategy Sheet.

To get the most out of a Requirement Intake Strategy Sheet, follow these best practices:

- Keep it simple: Ensure the form is easy to understand and complete.
- Regularly review and update: Review and update the form regularly to reflect changes in job requirements or industry trends.
- Use it consistently: Use the form consistently for all job requirements to ensure a standardized approach.

HOW TO ADJUST YOUR RECRUITMENT STRATEGY.

Recruitment is a dynamic process that requires flexibility and adaptability. As you navigate the recruitment landscape, you may encounter challenges or changes that necessitate adjustments to your strategy. Let's explore when and how to adjust your recruitment strategy.

1. After 1-2 Rejected Submittals

If you have submitted candidates and received rejection as feedback, it is essential to adjust your strategy to improve the quality of your submittals.

- Ask for feedback: Request feedback from the hiring manager or client to understand the reasons for rejection. This feedback can help you identify areas for improvement.
- Re-calibrate your Boolean search: Refine your Boolean search strings to better target candidates with the required skills and experience.
- Adjust your screening focus: Re-evaluate your screening criteria to ensure you are assessing the most relevant skills and qualifications.

2. After Multiple Interviews, No Shows

If you are experiencing a high rate of interview no-shows, it's time to adjust your strategy to improve candidate engagement and attendance.

- Introduce video pre-screens: Consider adding video pre-screens to the interview process. This can help candidates become more invested in the process and reduce the likelihood of no-shows.
- Use calendar confirmation tools: Implement calendar confirmation tools to remind candidates of upcoming interviews and reduce the risk of no-shows.

3. If Market Dries Up

If the market is drying up and you're struggling to find suitable candidates, it's essential to think creatively and explore alternative strategies.

- Proactively advise on alternates: Suggest alternative options, such as remote or part-time work arrangements, or consider candidates with slightly less experience.
- Expand your search: Consider expanding your search to include candidates from adjacent industries or with transferable skills.
- Re-evaluate your job requirements: Assess whether the job requirements are too rigid or if there is room for continuous changes.

To adjust your strategy effectively, Stay agile —that is, be willing to adapt your plan as needed. Communicate with clients by keeping them informed about changes to your strategy and the reasons behind them. Continuously evaluate and improve, that is, regularly assess your strategy's effectiveness and make adjustments to optimize results.

BUILDING CONSISTENCY IN DELIVERY WHILE STAYING FLEXIBLE IN EXECUTION.

In Recruitment, consistency, and flexibility are two essential elements that can make or break a team's success. Consistency ensures that clients receive high-quality candidates and services, while flexibility allows teams to adapt to changing client needs and market conditions. Let's explore how to build consistency in delivery while staying flexible in execution.

BUILDING CONSISTENCY.

Consistency is critical in Recruitment, as it ensures that clients receive a high level of service and quality candidates. To build consistency:

1. Establish transparent processes: Develop and document clear processes for sourcing, Screening, and submitting candidates. This helps ensure that all team members follow the same procedures, thereby reducing errors and inconsistencies.
2. Set standards: Establish standards for candidate quality, communication, and delivery. This helps ensure that clients receive a consistent level of service and quality candidates.
3. Provide training: Offer ongoing training and support to team members to ensure they possess the necessary skills and knowledge to deliver high-quality results consistently.
4. Monitor performance: Regularly monitor team performance and provide feedback to ensure that standards are being met.

STAYING FLEXIBLE.

Building Consistency in Delivery while Staying Flexible in Execution.

While consistency is essential, flexibility is equally important. Recruitment is a dynamic field, and teams must be able to adapt to shifting client needs and evolving market conditions. To stay flexible:

1. Be responsive: Be responsive to client needs and feedback. This helps ensure that clients receive the services they need and that the team is delivering value.
2. Adjust to changing market conditions: Stay current with market trends and adjust your strategies accordingly. This helps ensure that the team remains competitive and can adapt to changing client needs.
3. Be open to new ideas: Encourage team members to share new ideas and approaches. This helps foster a culture of innovation and continuous improvement.
4. Embrace technology: Leverage technology to streamline processes and improve efficiency. This helps teams stay flexible and adapt to changing client needs.

Balancing Consistency and Flexibility

Having learned about the two, we are going to talk about how to balance both of them and their importance as well:

1. Establish a framework: Create a structure that provides consistency while allowing for flexibility and adaptability.
2. Empower team members: Empower team members to make decisions and take ownership of their work. This helps foster a culture of accountability and flexibility.
3. Continuously evaluate and improve: Regularly assess processes and procedures and implement necessary improvements. This helps ensure that the team remains agile and adaptable.
4. Communicate effectively: Communicate clearly and concisely with clients and team members to ensure that everyone is aligned and informed about changes.

The Benefits are:

- Improved client satisfaction: Consistency and flexibility help ensure that clients receive high-quality services and candidates.
- Increased efficiency: Streamlined processes and flexibility can help reduce errors and improve efficiency.
- Competitive advantage: Teams that can strike a balance between consistency and flexibility are better positioned to adapt to changing

market conditions and client needs, thereby gaining a competitive advantage.

Case Example: A recruiter had an experience recently while she was sourcing for a Saas Implementation Consultant. She started sourcing candidates using LinkedIn only, but after a week, she had not received any candidates. This was because LinkedIn is limited in the sense that her target audience might not be using LinkedIn. She soon switched gears and introduced new sourcing channels, which are referrals and GitHub searches. Within three days, she successfully filled the role. This illustrates the importance of diversifying sourcing channels.

NOTE: This example is scenario-based for training and learning purposes to help us understand the scope better. The real scope can be much deeper and more precise.

Self Assessment

1. Why is a uniform strategy ineffective across roles?
2. Name two factors that can influence your recruitment plan.
3. What is one benefit of using a strategy intake sheet?
4. How might you adapt if candidates repeatedly reject interview invites?
5. What is the purpose of a strategy customization canvas?

Reflective Prompt: Think of a role you recently worked on. If you had to restart that search today, what would you do differently in your strategy?

CHAPTER FIFTEEN

DIVERSITY, EQUITY, AND INCLUSION (DEI) in the US HIRING SYSTEM

Objectives of this Chapter:

- Understand the role of DEI in US recruitment practices
- Learn key compliance principles from EEOC and related regulations
- Identify inclusive language and avoid biased terms in job descriptions
- Improve candidate experience through inclusive outreach and communication
- Build credibility with clients by demonstrating DEI awareness.

Before we go further, let us define these terms:

Diversity, equity, and inclusion are interconnected concepts that are essential for creating a fair and welcoming environment.

Diversity refers to the presence of different groups or individuals with unique characteristics, experiences, and perspectives within a given community or organization. This can include Demographic diversity, that is, differences in age, gender, ethnicity, sexual orientation, disability, and other characteristics. Cognitive diversity is differences in thought, perspective, and problem-solving approaches. And finally, Experiential diversity, which refers to differences in background, experience, and expertise.

Equity, on the other hand, refers to the fair treatment, access, and opportunities for all individuals, regardless of their background or characteristics. Equity involves fairness, which means making sure that policies and practices are fair and unbiased. Access means providing equal opportunities for resources, support, and access to resources. Opportunities, which means creating opportunities for growth, development, and advancement.

Inclusion refers to the creation of an environment where all individuals feel valued, respected, and supported. Inclusion fosters a sense of belonging among all individuals, encouraging participation and contribution from everyone and providing support and resources to help individuals thrive.

Why does Diversity, Equity, and Inclusion (DEI)Matter?

Diversity, equity, and inclusion (DEI) are essential components of any organization, community, or society. They foster a culture of respect, empathy, and understanding, resulting in numerous benefits. Let's talk about why DEI matters:

Benefits of Diversity.

1. Innovation and creativity: Diverse perspectives and experiences drive innovation and creativity. There's a saying that "Two heads are better than one." When everyone contributes their unique ideas, there will be more innovations.
2. Better decision-making: Diverse teams make more informed and well-rounded decisions.
3. Global competitiveness: Organizations with diverse workforces are better equipped to compete in a global market, as their candidates or staff come from different regions of the world.
4. Talent attraction and retention: Diverse organizations attract and retain top talent. Being diverse is a way of elevating the company's reputation, and highly talented candidates want to be in a company where they can meet others and learn from them.

Benefits of Equity.

1. Fairness and justice: Equity promotes fairness and justice, ensuring equal opportunities for all regardless of who they are or the position they hold. Note that it respects your status, but when there is a need for fairness, it is always fair.
2. Increased employee engagement: Employees who feel treated fairly are more engaged and motivated.
3. Improved reputation: Organizations that prioritize equity are seen as attractive employers and responsible corporate citizens. As a result, clients are more likely to want to work with them, candidates are more likely to want to be part of the company, and top investors are more likely to want to invest in them.

4. Better outcomes: Equity leads to improved outcomes, including enhanced health, education, and economic opportunities. Of course, when there is an enhanced reputation, there will be better outcomes.

Benefits of Inclusion

1. Sense of belonging: Inclusive environments create a sense of belonging among all individuals.
2. Increased collaboration: Inclusive teams work more effectively together, resulting in better outcomes. When there is no inclusion, a particular kind of tension arises, which can lead to poor collaboration and productivity.
3. Improved employee satisfaction: Employees who feel included are more satisfied and engaged, and they are even more likely to take on additional tasks because they are happy and satisfied.
4. Better representation: Inclusive organizations are more likely to represent diverse perspectives and experiences.

WHY DOES DIVERSITY, EQUITY, and INCLUSION (DEI) MATTER IM THE WORKPLACE?

We have discussed each of these concepts individually, but now we will learn about them collectively and explore why they are essential in the workplace.

Why DEI Matters in the Workplace:

1. Competitive advantage: Organizations that prioritize DEI are more likely to attract top talent, drive innovation, and improve decision-making.
2. Employee well-being well-being: DEI promotes employee well-being, reducing turnover and enhancing job satisfaction.
3. Reputation and brand: Organizations that prioritize DEI are seen as attractive employers and responsible corporate citizens.
4. Social responsibility: Prioritizing DEI is a social responsibility that promotes fairness, justice, and equality.

DIVERSITY, EQUITY AND INCLUSION (DEI) IN RECRUITMENT.

Diversity, Equity, and Inclusion (DEI) are no longer just buzzwords; they are essential components of the hiring landscape. Clients are increasingly expecting recruiters to understand and support diversity, equity, and inclusion (DEI) in their practices. DEI in Recruitment refers to ensuring that the hiring process is fair, inclusive, and diverse. It involves:

- Attracting candidates from diverse backgrounds
- Reducing bias in job descriptions, Screening, and interviews
- Creating an inclusive interview process
- Focusing on skills and qualifications rather than demographics
- Promoting equity and fairness in hiring decisions

The goal is to find the best candidate for the role while creating a workplace that values and respects diversity.

So, Why does DEI matter in Recruitment?

Diversity, Equity and Inclusion matters in Recruitment for the following reasons:

1. Business imperative: DEI is essential for businesses to attract top talent, foster innovation, and enhance decision-making within the organization.
2. Client expectations: Clients expect recruiters to prioritize Diversity, Equity, and Inclusion (DEI) in their practices, ensuring diverse candidate pipelines and inclusive hiring processes.
3. Reputation and brand: Recruiters who prioritize DEI are seen as attractive partners and contributors to a more equitable workplace.
4. Innovation and creativity: DEI drives innovation and creativity by bringing together diverse perspectives and experiences.
5. Better decision-making: DEI fosters more informed decision-making by considering diverse viewpoints and experiences.
6. Employee engagement and retention: DEI fosters employee engagement and retention by cultivating a sense of belonging and embracing diverse perspectives.
7. Social responsibility: DEI promotes social responsibility by fostering fairness, justice, and equality.

To support DEI in Recruitment, you can try the following strategies:

1. Inclusive job descriptions: Write job descriptions that are inclusive and appealing to diverse candidates.

2. Diverse candidate sourcing: Use diverse sourcing channels to attract candidates from underrepresented groups.
3. Bias-free Screening: Implement bias-free screening processes to ensure fair consideration of all candidates.
4. Culturally sensitive communication: Communicate with candidates in a culturally sensitive manner, respecting their backgrounds and experiences.

Ways to achieve DEI in Recruitment

1. Develop a DEI strategy: Develop a DEI strategy that aligns with client goals and priorities.
2. Train recruiters: Provide training and education on Diversity, Equity, and Inclusion (DEI) to recruiters, ensuring they understand its importance and best practices.
3. Use inclusive language: Use inclusive language in job descriptions, marketing materials, and communication with candidates.
4. Monitor and track metrics: Monitor and track DEI metrics, such as candidate demographics and hiring rates, to identify areas for improvement.

Benefits of Prioritizing DEI in Recruitment.

1. Access to diverse talent: Prioritizing DEI enables access to a diverse talent pool, driving innovation and business growth.
2. Improved reputation: Recruiters who prioritize DEI are seen as attractive partners and contributors to a more equitable workplace.
3. Increased client satisfaction: Clients are increasingly seeking recruiters who prioritize Diversity, Equity, and Inclusion (DEI), resulting in higher client satisfaction and loyalty.
4. Better business outcomes: DEI is linked to improved business outcomes, including enhanced decision-making, innovation, and financial performance.

KEY DEI CONCEPTS.

Several key DEI concepts are essential in any environment, whether it be work, education, or others. Let's see those concepts:

1. Diversity

Representation matters: Diversity refers to the representation of different identities, including:

- Race and ethnicity: Race and ethnicity refer to a person's identity based on their:

Physical characteristics(e.g., skin color, facial features) - Race

Cultural heritage, nationality, or ancestry (e.g., African American, Hispanic, Asian) - Ethnicity

- Gender and sexual orientation: - Gender is a person's internal identity as male, female, both, or neither (e.g., male, female, non-binary, transgender). Sexual orientation is a person's romantic or emotional attraction to others (e.g., straight, gay, lesbian, bisexual, pansexual).
- Disability: This is a physical, mental, or sensory impairment that can affect a person's ability to perform daily activities or interact with the world around them.
- Age: The length of time a person has lived, often categorized by stages such as youth, adult, or senior.

Other characteristics:

- Valuing differences: Diversity recognizes and values the unique experiences, perspectives, and contributions of individuals from different backgrounds.

2. Equity

Fairness is key: Equity refers to the principle of fairness in access, opportunity, and advancement. It involves:

- Identifying and addressing barriers: Removing obstacles that prevent individuals from accessing opportunities.
- Providing support and resources: Offering support and resources to help individuals succeed.
- Promoting equal opportunities: Ensuring equal opportunities for advancement and growth.

- Creating a level playing field: Equity aims to create a level playing field where everyone has an equal chance to succeed.

3. Inclusion

Actively embracing diverse voices: Inclusion involves actively incorporating diverse voices into decision-making and cultural processes. It involves:

- Creating a sense of belonging: Fostering a sense of belonging among all individuals.
- Encouraging participation: Encouraging participation and contribution from diverse individuals.
- Valuing diverse perspectives: Valuing diverse perspectives and experiences.
- Creating a culture of inclusion: Inclusion creates a culture where everyone feels valued, respected, and supported.

EEOC COMPLIANCE GUIDELINES.

The Equal Employment Opportunity Commission (EEOC) sets guidelines to ensure employers comply with federal laws prohibiting employment discrimination. The key aspects of EEOC include:

Protected Classes

The EEOC protects the following classes from employment discrimination:

1. Race
2. Gender
3. Religion
4. National origin
5. Age (40+)
6. Disability
7. Genetic information
8. Pregnancy

Compliance Requirements.

To comply with EEOC guidelines, employers must:

1. Avoid discriminatory language: Ensure job descriptions, screening questions, and interviews are free from bias or exclusionary language.

2. Use neutral language: Use neutral language in job postings and descriptions to attract diverse candidates.
3. Focus on essential duties: Focus on essential job duties and qualifications to avoid unintentional bias.
4. Train hiring managers: Train hiring managers and interviewers on EEOC compliance and avoiding bias.

Ways to achieve EEOC Compliance:

1. Regularly review policies: Regularly review policies and procedures to ensure compliance with EEOC guidelines.
2. Provide training: Provide training on EEOC compliance and diversity, equity, and inclusion.
3. Monitor hiring practices: Ensure fairness and equity in hiring practices.
4. Address complaints: Address complaints of discrimination promptly and thoroughly.

Consequences of Non-Compliance.

When you fail to comply with the EEOC guidelines, of course, there will be consequences for it. EEOC is a federal body, so you must abide by their laws if you want your company to stand or if you're going to last long in the recruitment industry. Failure to comply with EEOC guidelines can result in:

1. Lawsuits and fines: A lawsuit is a legal action taken against an individual or organization, typically to resolve a dispute or seek damages. Fines are monetary penalties imposed by a court or government agency as punishment for violating laws or regulations.
2. Reputation damage: This refers to the harm or negative impact on an individual's or organization's character, image, or standing, often resulting from negative publicity, scandals, or poor actions.
3. Loss of talent: This is when skilled or valuable employees leave an organization, often due to poor work environment, lack of opportunities, or dissatisfaction, resulting in a loss of expertise and potential future success.
4. Decreased employee morale: This refers to a decline in employees' overall attitude, satisfaction, and motivation, often resulting from poor management, a lack of recognition, or an unpleasant work environment, which leads to reduced productivity and engagement.

By following EEOC compliance guidelines, employers can ensure a fair and inclusive hiring process, reducing the risk of discrimination claims and promoting a positive work environment.

USING INCLUSIVE LANGUAGE IN JOB DESCRIPTIONS.

Job descriptions are often the first point of contact between a potential candidate and an employer. Using inclusive language in job descriptions can help attract a diverse pool of candidates and promote a positive employer brand. So, let's explore how to use inclusive language in job descriptions:

Preferred Qualifications vs. Must Haves

When writing job descriptions, it is crucial to distinguish between preferred qualifications and essential requirements.

- Preferred qualifications: These are skills or experiences that are nice to have but not essential for the role. Using "preferred qualifications" instead of "must have" can help attract a broader range of candidates.
- Must haves: These are essential skills or experiences required for the role. Use "must have" only when the qualification is genuinely crucial.

Using Gender-Neutral Terms

Using gender-neutral terms can help create a more inclusive job description.

- They instead of they: Using "they" instead of "he/she" can help create a more inclusive tone.
- Salesperson instead of salesman: Using gender-neutral job titles like "salesperson" instead of "salesman" can help attract candidates from diverse backgrounds.

Emphasis on Skills and Outcomes

When writing job descriptions, it is essential to focus on skills and outcomes rather than just credentials.

- Skills and outcomes: Emphasize the skills and outcomes required for the role instead of just listing credentials.
- Transferable skills: Consider highlighting transferable skills that can be applied to the role rather than just focusing on specific credentials.

When we say transferable skills, we mean skills or abilities that are valuable in multiple roles or industries, such as communication, problem-solving, leadership, time management, teamwork, etc.

Words and Phrases to Avoid.

Certain words and phrases can inadvertently deter candidates from applying. Here are some words and phrases to avoid:

- Age-coded words: Avoid using words like "young" or "energetic" that may inadvertently deter older candidates.
- Gendered phrases: Avoid using phrases like "ninja" or "rockstar" that may be perceived as masculine or feminine.
- Unnecessary requirements: Avoid requiring unnecessary degrees or US citizenship unless legally required.

Benefits of Inclusive Language.

Using inclusive language in job descriptions can have numerous benefits, including:

- Attracting diverse candidates: Inclusive language can help attract a diverse pool of candidates.
- Promoting a positive employer brand: Inclusive language can help promote a positive employer brand and demonstrate a commitment to diversity and inclusion.
- Reducing bias: Using inclusive language can help reduce bias in the hiring process.

Ways to write Inclusive Job Descriptions:

To write inclusive job descriptions, follow these steps:

- Use clear and concise language: Avoid using jargon or overly technical language that applicants might not be able to understand.
- Focus on essential duties: Focus on the essential duties and responsibilities of the role.
- Use inclusive language: Use inclusive language and avoid words or phrases that may inadvertently deter candidates.

- Regularly review and update: Review and update job descriptions regularly to ensure they remain inclusive and relevant to current needs.

By using inclusive language in job descriptions, employers can attract a diverse pool of candidates, promote a positive employer brand, and reduce bias in the hiring process.

INCLUSIVE VS NON-INCLUSIVE JOB LANGUAGE.

TOPIC	INCLUSIVE TERM	NON-INCLUSIVE EXAMPLE
Gender	Salesperson	Salesman
Age	Entry level opportunity	Recent graduate
Citizenship	Work authorization needed	US citizens only
Tone	Collaborative team	Fast-paced, high-pressure

PRACTICAL DEI ACTIONS FOR RECRUITERS.

As a recruiter, you play a crucial role in promoting Diversity, Equity, and Inclusion (DEI) in the hiring process. Here are some practical actions you can take to support DEI:

1. Ask Clients about DEI Goals or Preferences.

- Understand client priorities: Ask clients if they have diversity, equity, and inclusion (DEI) goals or preferences for the role or organization.
- Tailor your approach: Use this information to tailor your recruitment approach and ensure you are meeting the client's diversity, equity, and inclusion (DEI) needs.

2. Use Sourcing Tools with Anonymized or Confidential Resume Screening.

- Reduce bias: Utilize sourcing tools that enable confidential resume screening to minimize unconscious bias.
- Focus on skills: Focus on the candidate's skills and qualifications rather than their demographic information.

3. Present Balanced Candidate Slates.

- Diverse candidate pool: Present candidate slates that reflect a balance of:
  - Gender: Ensure a mix of male and female candidates.

- o Geography: Consider candidates from different locations.
  - o Background: Include candidates from diverse backgrounds and industries.
- Increase the chances of finding the best candidate: By presenting a diverse slate, you increase the likelihood of finding the ideal candidate for the role.

4. Educate Yourself on Implicit Bias

- Understand implicit bias: Educate yourself on implicit bias and its impact on the hiring process.
- Free resources and training: Utilize free resources and training to learn more about implicit bias and how to reduce it.
- Improve your skills: Continuously enhance your skills and knowledge to ensure you effectively promote diversity, equity, and inclusion (DEI) in the hiring process.

CASE EXAMPLE:

A recruiter submitted a strong candidate, but the job requirements said only US citizens could apply. The recruiter pointed out that this was not necessary since it wasn't a government job. The client realized that, too, and changed the requirements to include non-citizens with work permits (Green Card holders and EAD candidates). This helped find more qualified candidates.

NOTE: This example is scenario-based for training and learning purposes to help us understand the scope better. The real scope can be much deeper and more precise.

Self Assessment

1. What does DEI stand for?
2. Name one federal law that impacts US hiring practices related to DEI.
3. Give two examples of biased language often found in job descriptions.
4. Why is inclusive language important in recruiting?
5. How can you support DEI goals even if your client doesn't ask for it directly?

Reflective Prompt: Think about one change you can make today to be more inclusive in your job postings or candidate outreach.

CHAPTER SIXTEEN

HANDLING REJECTIONS, DROPOUTS AND COUNTEROFFERS

Objectives of this Chapter:

- Understand common reasons why candidates reject offers or drop out mid-process
- Learn how to coach candidates through counteroffer situations
- Identify early warning signs of disengagement
- Use frameworks to reduce dropout risk and improve closing rates
- Strengthen your ability to manage rejection with professionalism and resilience.

Just like we always do, before we start talking about Rejections, dropouts, and counteroffers, we have to know what they mean. So, below are definitions for each term:

1. Rejections: When a candidate is turned down or refused a job position, often due to not meeting qualifications or being outmatched by other candidates.
2. Dropouts: Candidates who withdraw or quit during the hiring process, often due to losing interest, accepting another offer, or personal reasons.
3. Counteroffer: A negotiation tactic where an employer offers improved terms (salary, benefits, etc.) to a candidate who has already received and is considering another job offer, aiming to retain or attract top talent.

Why Candidates Say No (and How to Prevent It) Even the best-matched candidates can say "no" to an offer. Sometimes, it's money, timing, or fear of change. At other times, it's unclear communication or a lack of trust. As a recruiter, you must learn how to uncover hesitations early, align expectations, and keep engagement high through every stage. In this chapter, we will discuss this topic and explore strategies to ensure candidates do not reject or reject those who refuse.

COMMON DROPOUT AND REJECTION TRIGGERS.

1. Unclear Job Details or Shifting Expectations.

Imagine you are excited about a job, but you are not sure what the role entails or what your responsibilities will be. This lack of clarity can lead to confusion and mistrust. If the employer changes the job requirements or expectations midway through the hiring process, it can be frustrating for the candidate.

Example: A candidate is told they will be working on project A, but during the interview process, the employer also mentions projects B and C. The candidate might feel uncertain about their role and responsibilities and may end up rejecting the job or drop out halfway through.

How to prevent it: Employers should clearly define job requirements and expectations from the start. They should also communicate any changes promptly and transparently.

2. Poor Candidate Experience or Delays in Feedback.

A poor candidate experience can be a major turn-off. This includes things like:

- Long wait times for feedback or responses
- Unprofessional communication or behavior from the hiring team
- Lack of updates on the hiring process

Example: A candidate attends an interview and doesn't hear back for weeks. When they finally receive feedback, it is often vague or unhelpful, and most of the time, the candidate cannot wait that long and has likely secured another job during the waiting period.

How to prevent it: Employers should prioritize communication and keep candidates informed throughout the hiring process. They should also ensure that all interactions with candidates are professional and respectful.

3. Better Offers (Compensation, Location, WFH Flexibility)

Sometimes, candidates receive better offers from other companies that outweigh the original job offer. This can include things like:

- Higher salary or benefits
- Better work-life balance or flexibility

- More attractive company culture or values

Example: A candidate receives a job offer with a higher salary and flexible work-from-home arrangements, making the original job offer less appealing. Of course, he would choose the new job offer, as it would provide him with a higher income and be less stressful as well.

How to prevent it: Employers should ensure their job offers are competitive and appealing. They should also consider offering benefits and perks that are important to candidates.

4. Counteroffers from Current Employer

A counteroffer occurs when a candidate's current employer presents a better deal to retain the candidate's services. This can include things like:

- A salary increase
- New responsibilities or opportunities
- Improved benefits or perks

Example: A candidate is about to leave their current job for a new opportunity, but their employer offers them a promotion and raises to stay, which is more attractive than the offer from the job he was about to switch to. He would change his mind and remain at his current job.

How to prevent it: Employers should understand that counteroffers are a regular part of the hiring process. They can address the candidate's concerns and needs before making a job offer.

5. Cold Feet or Family Pressure

Sometimes, candidates get cold feet or face pressure from family or friends about their job decision. This can include things like:

- Fear of change or uncertainty
- Concerns about company culture or values
- Family or friends advising against the job

Example: A candidate is hesitant to take a job in a new city because they are worried about leaving their family behind. Or, he is reluctant because a friend had told him a few things about that particular job. All these would make him want to lose interest in the job and look for something else.

How to prevent it: Employers should provide candidates with clear information about the company culture and values. They can also offer support and resources to help candidates transition smoothly. They can also ease the candidate's fears if they ever discover it.

By understanding these standard dropout and rejection triggers, employers can take proactive steps to address candidate concerns and increase the chances of a successful hire.

NOTE: All examples in this context are scenario-based for training and learning purposes to help us understand the scope better. The real scope can be much deeper and more precise.

COUNTEROFFER COACHING.

What is a Counteroffer?

A counteroffer is when a candidate's current employer offers them a better deal to stay with the company after they've decided to leave. This can include things such as a salary increase, promotion, or enhanced benefits.

Why is Counteroffer Coaching Important?

Counteroffer coaching is essential because it helps candidates prepare for potential counteroffers and make informed decisions about their careers.

How to Prepare Candidates:

1. Discuss Counteroffers Before the Offer Stage

It is crucial to discuss counteroffers with candidates before extending a job offer. This conversation can help candidates think critically about their decision to leave their current employer.

Example Question: "What will you do if your current employer makes a counteroffer to keep you?"

This question helps candidates reflect on their priorities and what they aspire to achieve in their careers. It would prompt them to reconsider if they love their current job but are dissatisfied with the payor if if they do not want to work there again at all.

2. Remind Candidates of Their Original Motivation for Change

Candidates often have reasons for leaving their current employer, such as limited growth opportunities, poor company culture, or inadequate compensation. Reminding them of these reasons can help them stay focused on their long-term goals.

Example: "You mentioned earlier that you are looking for new challenges and growth opportunities. Keep that in mind when considering any counteroffers."

3. Share Statistics: Over 50% of Those Who Accept Counteroffers Leave Within a Year

Sharing statistics about counteroffers can help candidates understand the potential risks of accepting a counteroffer. Research shows that many employees who accept counteroffers ultimately leave their company within a year due to various dissatisfactions, as the allure is often fleeting, typically lasting only the first two to three months.

Example: "Did you know that over 50% of employees who accept counteroffers leave their company within a year? Let's consider whether a counteroffer is truly in your best interest."

4. Encourage Transparency and Focus on Long Term Growth

It is essential to encourage candidates to be transparent with their current employer about their decision to leave. Additionally, reminding them to focus on long-term growth and career goals can help them make a more informed decision.

Example: "Remember, it's not just about short-term gains. Consider what's best for your long-term career growth and goals."

Benefits of Counteroffer Coaching:

1. Informed Decision Making: Candidates are better equipped to make informed decisions about their careers.
2. Reduced Risk: Candidates are less likely to accept a countcroffer that might not be in their best interest.
3. Increased Job Satisfaction: Candidates are more likely to be satisfied with their new role and employer.

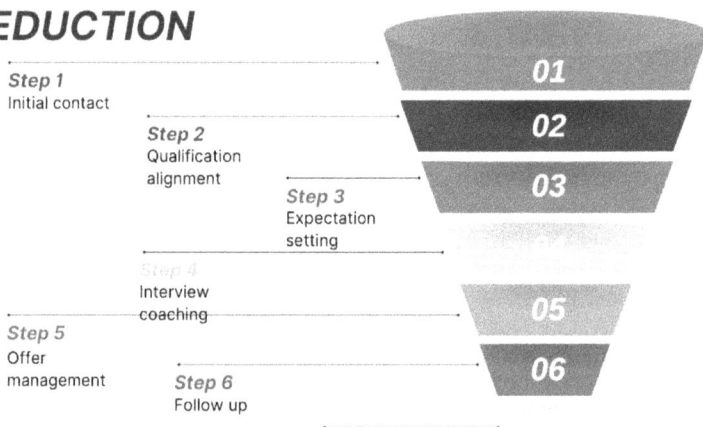

DROP OUT RISK REDUCTION

Step 1
Initial contact — 01

Step 2
Qualification alignment — 02

Step 3
Expectation setting — 03

Step 4
Interview coaching — 05

Step 5
Offer management

Step 6
Follow up — 06

RED FLAGS THAT SIGNAL POSSIBLE DROPOUTS.

These are things a recruiter should look out for in a candidate that could indicate if the candidate would drop out along the line. Some of them are:

1. Slow or Vague Communication

When a candidate is slow to respond or their communication is vague, it can be a sign that they are losing interest or busy with other priorities.

Examples:

- Not responding to emails or messages promptly
- Giving brief or unclear answers to questions
- Not following up on scheduled interviews or meetings

What it might mean: The candidate might be exploring other opportunities or has lost enthusiasm for the role.

How to address it: Send a gentle reminder or clarify expectations. If there is still no response from the candidate, it might be time to move on.

2. No Clear Availability for Interviews

When a candidate has trouble scheduling interviews or doesn't provide clear availability, it can indicate a lack of commitment or prioritization.

Examples:

- Constantly rescheduling interviews
- Not providing specific dates or times for interviews
- Being unavailable for extended periods

What it might mean: The candidate might be too busy or not interested enough to make time for the interview process.

How to address it: Try to schedule interviews well in advance or offer flexible scheduling options. If the candidate continues to struggle with availability, it might be a sign that they are not a good fit.

3. Hesitation Around Salary/Rate Discussions

When a candidate is hesitant to discuss salary or rates, it can indicate uncertainty or unrealistic expectations.

Examples:

- Avoiding discussions about compensation
- Being vague about salary expectations
- Seeming uncomfortable or evasive when discussing pay

What it might mean: The candidate might be unsure about their worth or have expectations that don't align with the company's budget.

How to address it: Have an open and transparent conversation about compensation. Provide clear information about the company's budget and expectations.

4. Asking Repetitive Questions Already Clarified

When a candidate asks repetitive questions that have already been answered, it can indicate a lack of attention or interest, as it is common for people to forget things they are not interested in easily.

Examples:

- Asking about job responsibilities or company culture when it has already been discussed
- Requesting information that is readily available on the company website or job description

- Seeming uninformed about the company or role despite previous discussions

What it might mean: The candidate might not be doing their research or is not genuinely interested in the role.

How to address it: Politely point out that the question has already been answered and offer additional information if needed. If the candidate continues to ask repetitive questions, it might be a sign that they are not a good fit.

Note: All points here had the modal verb "might," which means that it might not always be the case.

What to Do When You Notice Red Flags:

1. Communicate openly: Address the issue directly and clarify expectations.
2. Assess the candidate's interest: Determine whether the candidate remains interested in the role and is willing to move forward.
3. Evaluate the candidate's fit: Consider whether the candidate's skills, experience, and attitude align with the company's needs and culture.
4. Move on if necessary: If the red flags persist and the candidate doesn't seem like a good fit, it might be time to explore other candidates.

By recognizing these red flags, you can save time and resources by focusing on candidates who are genuinely interested and qualified for the role.

HOW TO HANDLE REJECTIONS PROFESSIONALLY

1. Thank the Candidate for Their Time and Honesty

When a candidate rejects your job offer or decides not to move forward in the hiring process, it is essential to thank them for their time and honesty.

Why? Being respectful and appreciative shows that you value their time and consideration, as well as the fact that they took the time to speak with you, which some other candidates might not do.

Example: "Thank you for letting us know about your decision. We appreciate your honesty and the time you took to interview with us."

2. Ask for Candid Feedback on What Didn't Feel Right

Asking for feedback can provide valuable insights into what didn't quite fit for the candidate. This can help you identify areas for improvement to benefit other candidates.

Why? Feedback can help you refine your hiring process, job description, or company culture.

Example: "Can you share with us what didn't feel right about the opportunity or our company? Your feedback is invaluable to us."

3. Maintain the Relationship (They May Return or Refer Others to you)

Even if a candidate is not a good fit for the current role, they might be interested in future opportunities or know someone who is.

Why? Maintaining a positive relationship can lead to future applications or referrals.

Example: "We would love to stay in touch and keep you updated about future opportunities that might be a better fit for you."

4. Share Lessons with Your Team to Refine Future Approach

Sharing feedback and lessons learned with your team can help refine your hiring approach and improve future candidate experiences. If the candidate gives feedback on what didn't feel right, you can discuss it with your team and then find a way around it.

Why? Continuous improvement is key to attracting top talent and building a strong employer brand.

Example: "Let's discuss the feedback we received and see how we can apply it to future hiring processes."

Benefits of Handling Rejections Professionally:

1. Positive Employer Brand: Candidates are more likely to speak positively about your company, even if they were not eventually hired.
2. Future Opportunities: Maintaining relationships can lead to future applications or referrals.

3. Improved Hiring Process: Feedback can help refine your hiring approach and enhance the candidate experience.
4. Professional Reputation: Handling rejections professionally demonstrates your company's values and commitment to respect and empathy.

Tips for Handling Rejections:

1. Respond promptly: Reply to candidates promptly, even if it's just to acknowledge their decision.
2. Be genuine: Show appreciation and empathy in your communication.
3. Keep it concise: Keep your messages brief and to the point.
4. Follow-up: If a candidate expresses interest in future opportunities, ensure that you follow up and keep them informed.

CASE EXAMPLE:

- A recruiter presented a job offer to a candidate.
- The candidate's current employer made a surprise counteroffer to keep them.
- The candidate accepted the counteroffer, and the recruiter lost the candidate.

So, what did the recruiter do about this? Or how did she react to her next candidate?

- The recruiter learned from the previous experience and decided to coach the next candidate about potential counteroffers from the start.
- The recruiter prepared the candidate for this possibility and discussed how to handle it.

And what was the outcome?

- When the same situation arose (the candidate's current employer made a counteroffer), the candidate was prepared and knew what to expect.
- The candidate declined the counteroffer and accepted the new job offer.
- The recruiter's proactive approach built long-term trust with the client, demonstrating their expertise and ability to navigate complex hiring situations.

NOTE: This example is scenario-based for training and learning purposes to help us understand the scope better. The real scope can be much deeper and more precise.

Self Assessment

1. Name two common reasons candidates reject job offers.
2. What question should you ask early to prepare for counteroffers?
3. What is one red flag that a candidate may drop out?
4. Why is it important to keep relationships intact after a rejection?
5. What's one tactic to reduce dropout risk during the interview phase?

Reflective Prompt: Think of a time a candidate backed out or declined at the last moment. What would you change in your process now to either prevent that or handle it better?

CHAPTER SEVENTEEN

BRANDING, TRUST AND DELIVERY

Objectives of this Chapter:

- Understand the long-term value of trust and delivery in recruitment
- Learn how personal branding builds recruiter credibility
- Explore how consistent delivery leads to client and candidate advocacy
- Adopt practices that strengthen your professional reputation
- Use a trust-building framework to evaluate and improve your brand presence.

In recruitment, your reputation is everything. It's like your brand is your currency, and it is what you use to buy trust and opportunities. When working with candidates or hiring managers, it's not just about filling a role; it's about building relationships and establishing trust.

Think about it like this: when you are dealing with someone for the first time, they don't just judge you based on who you are today; they also judge you based on what they have heard about you in the past. Your past interactions, delivery track record, and communication style all contribute to your perceived value. That is precisely how it is for all forms of businesses, recruiters, and hiring managers. If a friend tells you that a restaurant sells horrible dishes, trust me, you would not want to eat there.

Great recruiters do not just focus on filling roles; they focus on building trust with their candidates and clients. They understand that recruitment is not just about finding the right person for the job but about creating a positive experience for everyone involved.

When you build trust, you create a reputation that precedes you. People start to see you as a reliable and competent recruiter who delivers results. This reputation can open doors and create opportunities that might not have been available otherwise.

Consistency is key to building trust and establishing a strong reputation. It is about consistently delivering high-quality candidates, communicating effectively, and following up on promises. When you are consistent, people know what to expect from you, and they are more likely to trust you.

In the long run, your reputation and consistency can make or break your career as a recruiter. It is worth investing time and effort into building strong relationships and delivering exceptional results. When you do, you will find that opportunities start to come to you, and you will be able to create a strong network of contacts that can help you succeed. In this chapter, this is what we are going to be talking about.

But, just like we always do, we are going to define the terms first.

Branding:

Branding refers to the unique image, reputation, and identity of a person, company, or organization. It is the way people perceive and remember you based on your values, personality, and overall experience. Branding encompasses elements such as logos, messaging, and visual identity, but it is more than just appearance; it is about the feelings and emotions people associate with you.

Trust:

Trust is the confidence and faith that someone has in another person, company, or organization. Trust is built when you consistently demonstrate reliability, honesty, and integrity. When people trust you, they believe you will do what you say you will do, and they will be more likely to work with you, recommend you, or support you.

Delivery:

Delivery refers to the act of fulfilling promises, meeting expectations, or providing results. In the context of recruitment, delivery might mean finding the right candidate for a job, meeting deadlines, or providing excellent service to clients. When you deliver on your promises, you build trust and credibility with others, and your reputation grows.

These three concepts are interconnected: strong branding can help build trust, and trust is built through consistent delivery. When you deliver on your promises, you reinforce your brand and build trust with others.

WHAT IS PERSONAL BRANDING IN RECRUITMENT?

Personal branding is the image or reputation that people associate with you, your name, and your work. In recruitment, it is how candidates, clients, and colleagues perceive you as a recruiter. Your brand is built through your actions, behavior, and communication style.

How is Personal Branding Built?

Your brand is built through various aspects of your professional life, including:

- Emails and Communication: How you write emails, respond to messages, and communicate with others. Is your tone friendly and approachable, or formal and professional? Do you know how to spell correctly? (Of course, no recruiter wants to be seen as one who is terrible at spelling or has bad grammar).
- Follow up and Follow through: How you follow up with candidates, clients, and colleagues. Do you keep your promises and meet deadlines? Do you get feedback? Do you help your clients follow up with a particular candidate they are interested in?
- Negotiation and Presentation: How you negotiate with clients and present candidates. Are you confident, knowledgeable, and persuasive? Are you lackadaisical? Do you have strong negotiation willpower willpower? Or are you easy to override?
- Online Presence: Your LinkedIn profile, online content, and social media presence. What kind of content do you share? Are your profiles up to date and professional?
- Recommendations and Referrals: What others say about you and your work. Do you receive glowing recommendations and referrals from clients and candidates? Do you receive unpleasant comments from others behind your back?
- Calendar Etiquette: How you manage your schedule and meetings. Are you punctual, respectful, and considerate of others' time?
- -Personal Character: How you act and behave. Do you have a good character? Are you respectful to your candidates? Do you try to reason with them? Do you understand things and carefully explain to them when they do not understand?

All of these are things to look at and questions to ask yourself about your branding.

Why is Personal Branding Important in Recruitment?

Your brand is essential in recruitment because it:

- Builds Trust: A strong personal brand helps build trust with candidates, clients, and colleagues. When people trust you, they are more likely to work with you and recommend you to others.
- Differentiates You: A unique and consistent personal brand sets you apart from other recruiters. It helps you stand out and establishes your identity in the industry.
- Increases Credibility: A professional and reliable personal brand increases your credibility and expertise in the eyes of others. It demonstrates that you are committed to your work and take care of your reputation.
- Opens Doors: A strong personal brand can open doors to new opportunities, relationships, and business growth. When people know and respect your brand, they are more likely to introduce you to new contacts and opportunities.

How to Build a Strong Personal Brand in Recruitment:

1. Be Consistent: Consistency is key to building a strong personal brand. Ensure that your communication style, behavior, and online presence are consistent across all platforms.
2. Be Authentic: Authenticity is essential to building trust and credibility. Be true to yourself and your values, and don't try to be someone you are not.
3. Provide Value: Focus on providing value to candidates, clients, and colleagues. This can be through helpful content, expert advice, or exceptional service.
4. Engage with Others: Engage with others on social media, attend industry events, and participate in online communities. This helps build relationships and establishes your presence in the industry.

By building a strong personal brand, you can establish yourself as a trusted and respected recruiter, attract new opportunities, and grow your business. You can also obtain candidates from others due to your trustworthiness.

ACTIONS THAT BUILD TRUST AND ENHANCE THE BRAND.

These are actions or steps that you must take to build trust and enhance your brand. If others do not trust your brand and you are not making any move to make them trust you, then there is no need to own that brand. The fact that you are nonchalant about people not trusting your brand already says a lot about who you are. So, here are ways to make people trust your brand more.

1. Always Follow Through

When you say you will do something, do it. If you promise to send information, then send it promptly. If you commit to a deadline, meet it. Following through on your commitments shows that you are reliable and responsible.

Why it matters: When you consistently follow through, people start to trust you. They know that you mean what you say and will do what you promise. This builds confidence in your abilities and strengthens your relationships.

2. Use Clear, Timely, and Respectful Communication

Communicate in a clear, easy-to-understand, and respectful manner. Be prompt in your responses and ensure your messages are concise and relevant.

Why it matters: Good communication helps avoid misunderstandings and shows that you value the other person's time and opinions. When you communicate effectively, people are more likely to trust and respect you.

3. Own Mistakes and Offer Quick Resolutions

If you make a mistake, own up to it and apologize. Offer a solution or a way to fix the problem, and do it quickly. If it is a problem that cannot be immediately fixed, then you should find a way to mitigate it.

Why it matters: When you take responsibility for your mistakes and offer solutions, you show that you are accountable and committed to doing things right. This builds trust and shows that you are proactive in resolving issues.

4. Highlight Your Process

Explain how and why you do things a certain way. Share your process for screening, qualifying, or recommending candidates.

Why it matters: When you share your process, you show that you are transparent and methodical in your approach. This helps build confidence in your abilities and shows that you are thorough and professional.

5. Share Hiring Success Stories (with consent)

Share stories of the successful hires you have made, with the candidate's consent. This shows your track record and demonstrates your expertise.

Why it matters: When you share success stories, you provide social proof and demonstrate your ability to deliver results. This builds credibility and trust with potential clients and candidates.

Benefits of These Actions:

1. Increased Trust: By following through, communicating effectively, owning mistakes, highlighting your process, and sharing success stories, you build trust with clients and candidates.
2. Enhanced Brand: These actions enhance your brand by showcasing your reliability, expertise, and commitment to delivering results.
3. Stronger Relationships: By being transparent, accountable, and communicative, you build stronger relationships with clients and candidates.
4. Improved Credibility: When you demonstrate your expertise and track record, you increase your credibility and establish yourself as a trusted authority in your field.

DIAGRAM.

The Flywheel Effect in recruitment branding refers to the momentum and growth that occur when a recruiter or recruitment agency consistently delivers high-quality candidates, builds strong relationships, and provides excellent service.

Imagine a flywheel: initially, it takes effort to get it spinning, but as it gains momentum, it becomes harder to stop. In recruitment branding, the flywheel effect works similarly:

1. Initial Effort: Building a strong reputation and brand takes time and effort.

2. Momentum Builds: As you consistently deliver quality results, your reputation grows, and more clients and candidates want to work with you.
3. Growth Accelerates: With increased momentum, your brand becomes more visible, and opportunities expand.

The flywheel effect creates a self-reinforcing cycle, as shown below.

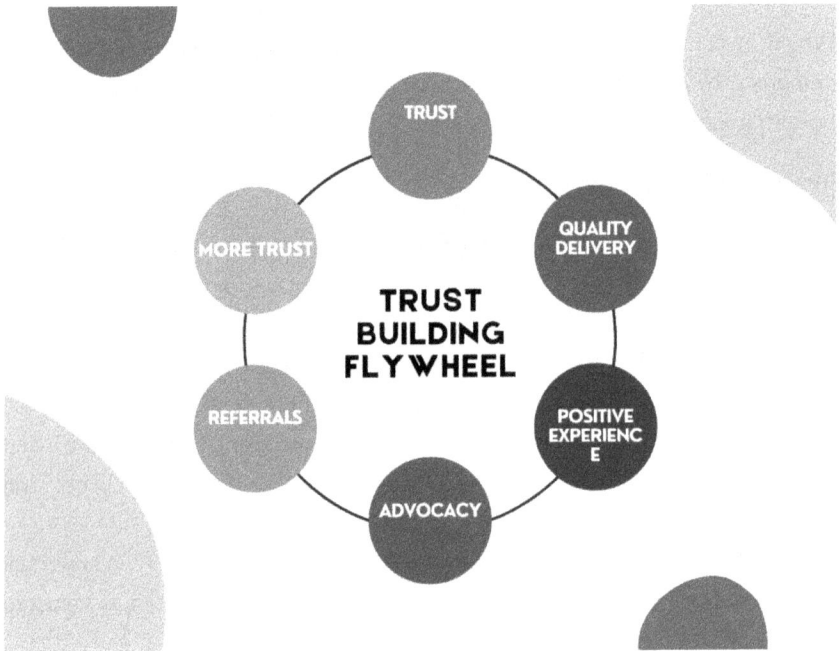

POSITIONING YOURSELF AS A GLOBAL TALENT PARTNER.

What does it mean to be a Global Talent Partner?

Being a Global Talent Partner means you are not just a recruiter but a trusted advisor who helps clients find the best talent globally. You are an expert in the market, and you understand the trends, challenges, and opportunities that come with finding top talent.

How to Position Yourself as a Global Talent Partner:

1. Speak with Confidence about Market Trends, Compensation Norms, and Process Timelines

When you speak with clients, show that you are knowledgeable about the market. Do not act like a complete novice, even if you are. Share insights on trends, compensation norms, and process timelines. This demonstrates your expertise and helps clients trust your judgment.

Example: "Based on my research, the current market trend for this role is a salary range of $X to $Y. I think we can find a great candidate within this range of pay."

2. Present Candidates like You are Pitching a Solution, Not Just Forwarding a Resume.

When presenting candidates, don't just send their resumes and expect the client to do all the work. Instead, pitch the candidate like you are selling a solution to a problem. Highlight their skills, experience, and achievements, and explain why they would be an excellent fit for the role.

Example: "I have found a candidate with exactly the skills and experience you are looking for. Let me tell you more about their background and why I think they would be a great fit for this role."

Instead of the word "think," you could use "know" to increase your confidence level.

3. Educate Your Clients: "Here's Why This Candidate Fits Beyond the Keywords"

Do not just focus on keywords and qualifications. Take the time to explain why a candidate is a good fit for the role beyond just their technical skills. Share insights on their experience, personality, and achievements, and demonstrate how they can contribute to the client's organization.

Example: "While the candidate's resume shows they have the technical skills for the role, I've also spoken with them about their experience working in similar environments and their passion for innovation. I think they will bring a fresh perspective to your team."

Benefits of Positioning Yourself as a Global Talent Partner:

1. Increased Credibility: By demonstrating your expertise and knowledge, you build credibility with clients and establish yourself as a trusted advisor.

2. Stronger Relationships: By taking the time to understand clients' needs and presenting solutions, you build stronger relationships and increase client loyalty.
3. Better Matches: By educating clients on the candidate's fit beyond keywords, you increase the chances of making better matches and reducing turnover.
4. Competitive Advantage: By positioning yourself as a Global Talent Partner, you differentiate yourself from other recruiters and establish a competitive advantage in the market.

Some Tips for Success:

1. Stay Up to Date: Stay current with market trends, compensation norms, and industry developments to maintain your expertise.
2. Understand Client Needs: Take the time to understand clients' needs and challenges and tailor your approach to meet their specific requirements.
3. Develop Your Skills: Continuously develop your skills and knowledge to stay ahead of the competition and provide exceptional service to clients.

PRACTICES THAT WOULD STRENGTHEN YOUR PROFESSIONAL REPUTATION.

What is a Professional Reputation?

Your professional reputation is the perception that others have of you as a professional. It is based on your actions, behavior, and performance in your work. A strong professional reputation can open doors to new opportunities, build trust with clients and colleagues, and increase your credibility in your industry.

Practices that Strengthen Your Professional Reputation:

1. Be Reliable and Dependable: Do what you say you will do, and do it on time. This shows that you are responsible and can be trusted to deliver results.
2. Communicate Effectively: Communicate concisely and respectfully with others. This helps to avoid misunderstandings and builds trust.

3. Be Proactive: Take initiative and anticipate potential problems or opportunities. This shows that you're proactive and committed to finding solutions.
4. Continuously Learn and Improve: Stay current with industry developments and best practices to enhance your skills and knowledge. Continuously learn and improve your skills and expertise to stay ahead of the competition.
5. Be Transparent and Honest: Be open and honest in your dealings with others. This builds trust and credibility, showing that you're committed to doing things right.
6. Show Appreciation and Gratitude: Express gratitude and appreciation to others who help you or support you. This shows that you value their contributions and care about their well-being.
7. Maintain a Professional Online Presence: Ensure that your online presence, including social media profiles and other digital platforms, is professional and consistent with your brand.
8. Deliver High-Quality Work: Consistently deliver high-quality work that meets or exceeds expectations. This builds trust and credibility with clients and colleagues.
9. Take Ownership and Accountability: Accept responsibility for your mistakes and actions. Be accountable for your work and decisions, and be willing to learn from your mistakes.
10. Show Empathy and Understanding: Show empathy and understanding towards others, including clients, colleagues, and candidates. This fosters strong relationships and demonstrates that you genuinely care about their needs and concerns.

Benefits of Strengthening Your Professional Reputation:

These benefits are similar to that of personal branding. They are both identical, but they have differences that we will discuss later. The benefits of strengthening your professional reputation include:

1. Increased Trust: A strong professional reputation builds trust with clients, colleagues, and other stakeholders
2. Improved Credibility. A strong professional reputation increases your credibility and establishes you as an expert in your field.

3. New Opportunities: A strong professional reputation can open doors to new opportunities, including new clients, projects, and career advancement.
4. Stronger Relationships: A strong professional reputation helps build stronger relationships with clients, colleagues, and other stakeholders.
5. Competitive Advantage: A strong professional reputation can be a competitive advantage in your industry, setting you apart from others and establishing you as a trusted and respected professional.

Tips for Strengthening Your Professional Reputation:

1. Set High Standards: Set high standards for yourself and strive to meet or exceed them. Nothing feels better than setting standards for yourself and exceeding them.
2. Be Consistent: Consistency is key to building a strong professional reputation. Consistently deliver high-quality work and behave professionally.
3. Seek Feedback: Seek feedback from others and use it as an opportunity to learn and improve.
4. Stay Up to Date: Stay current with industry developments and best practices to maintain your expertise and stay ahead of the competition.
5. Be Authentic: Be authentic and genuine in your interactions with others. This builds trust and credibility, showing that you are committed to being true to yourself and others.

By following these practices and tips, you can strengthen your professional reputation, build trust and credibility, and establish yourself as a respected and trusted professional in your industry.

DIFFERENCES BETWEEN PERSONAL BRANDING AND PROFESSIONAL REPUTATION.

Personal Branding and Professional Reputation are similar but different concepts:

Personal Branding:

- Refers to the image, values, and personality you present to the world
- Encompasses your unique identity, values, and story
- Often intentional and curated (e.g., social media profiles, personal website)

- Focuses on showcasing your personality, skills, and expertise

Professional Reputation:

- Refers to the perception others have of your work, behavior, and expertise
- Based on your actions, performance, and interactions with others in a professional context
- Can be influenced by both intentional and unintentional factors (e.g., work quality, communication style, online presence)
- Focuses on demonstrating your competence, reliability, and trustworthiness

Key differences:

1. Intentionality: Personal branding is often intentional, while both deliberate and unintentional factors can impact a professional's reputation.
2. Focus: Personal branding emphasizes showcasing your personality and identity, whereas professional reputation focuses on demonstrating your competence and trustworthiness.
3. Scope: Personal branding can extend beyond professional contexts, while professional reputation is primarily concerned with your work and professional interactions.

While personal branding and professional reputation are distinct, they can overlap and influence one another. A strong personal brand can contribute to a positive professional reputation, and a good professional reputation can enhance your brand. It's vice versa.

CASE EXAMPLE:

A recruiter worked with a client and built a good reputation by sending only a few highly qualified candidates for job openings. Instead of sending numerous resumes, the recruiter prioritized quality over quantity.

The client became impressed with the recruiter's work and began sending all the most important job openings directly to them. The client even skipped working with other recruiters or vendors.

Do you know why this worked?

It worked because the recruiter's approach demonstrated that they understood the client's needs and were focused on delivering high-quality candidates. This built trust and credibility, making the client want to work directly with the recruiter.

NOTE: This example is scenario-based for training and learning purposes to help us understand the scope better. The real scope can be much deeper and more precise.

Self Assessment

1. What are three elements that contribute to your personal recruiter brand?
2. Why is delivery consistency important in building client trust?
3. How can sharing your process enhance credibility?
4. What is one outcome of building strong trust with candidates?
5. What is the flywheel effect in recruitment branding?

Reflective Prompt: Think of someone you trust professionally. What specific behaviors or habits made you trust them? How can you adopt those in your recruitment practice?

CHAPTER EIGHTEEN

CAREER GROWTH AND PROFESSIONAL DEVELOPMENT

Objectives of this Chapter:

- Explore career progression paths in the global recruitment industry
- Understand what skills and experiences help transition from recruiter to leader
- Learn about certifications, courses, and communities that support growth
- Discover how to create a personal development roadmap
- Build habits that support long-term success and professional identity.

Once you gain a deeper insight into this industry, you will see that we often move quickly, placing people into roles one after another. However, as recruiters, we have the power to design careers, not just put people in positions.

Whether your goal is to move up to leadership, get specialization in one specific area, or become a strategic partner to your clients, it's fundamentally important to invest in yourself first. Having a successful career doesn't just happen organically; it occurs at the right time. It should be something you work towards every single day and not something you should settle for.

Think of what career you want to go after. Which areas of your personality are most promising? What skills/knowledge do you need to learn for your goals? If you are intentional about the changes you make in your career, you will be able to break even faster and meet your career goals.

Investing in yourself is more than simply taking courses or attending training sessions; it's about being an active participant, looking for opportunities, taking advantage of your experiences, cultivating connections, seeking mentorship,p, and keeping current with the industry.

When you prioritize your career development, you'll become a more effective recruiter. You'll be able to offer more value to both your clients and candidates, build stronger relationships, make better matches, and achieve better outcomes.

So, we encourage you to take ownership of your career development - truly CAREER, be intentional about where you go, and invest in yourself, whether big or small. You always have room to grow and improve.

By doing so, you will not only reach your career goals, but you will also become a more confident, competent, and booming recruiter while building a career that you enjoy and making a positive impact on other people's lives.

Before we go further, let's go ahead and define the terms "career growth" and "professional development."

Career Growth The pursuit and progression of an individual's career over time, based on new tasks, responsibilities, and new experiences, which generates greater satisfaction, recognition, and compensation, either vertically (facing a superior level in an organization or laterally) or in a specific area of interest.

Professional Development: In contrast, this refers to learning, training, and self-development that enhance one's ability to perform professionally and increase opportunities for career success. It includes ongoing learning, continuing education, and self-development to stay current with industry developments, best practices, and technological advances.

Key Aspects of Career growth and Professional development:

1. Learning: Acquiring new skills, knowledge, and competencies.
2. Self Improvement: Enhancing performance, productivity, and effectiveness.
3. Career Advancement - Preparing yourself for promotions, new positions, or other transitions in your career.
4. Networking: Establishing relationships and contacts within one's field or profession.

Benefits of Career growth and Professional development:

1. Improved job satisfaction: Increased skills and knowledge can lead to increased confidence and job satisfaction.
2. Performance Improvement: Professional Development Improves Performance, Productivity, and Effectiveness.
3. Professional Development and Career Advancement: Investing in professional development can result in promotions, new roles, and opportunities opportunities.

4. Competitive Advantage: Knowledge of industry trends and best practices helps professionals stand out in the job market.

GROWTH MILESTONES: From Recruiter to Talent Leader.

Beginner (Entry Level):

At this stage, you are more focused on learning the basics of recruitment. Your responsibilities might include:

- Job Boards: Posting job ads on various job boards and managing job postings.
- Resume Sourcing: Finding potential candidates by searching through resumes and databases.
- Basic Screening: Review resumes and cover letters to determine if candidates meet the minimum qualifications for a role.

As a beginner, you are just building your foundation in recruitment. You are learning about the industry, developing your skills, and getting familiar with the tools and technologies used in recruitment. Therefore, it is only right that you start with straightforward tasks, and someone must be watching you to ensure you do not make serious mistakes.

Intermediate (Middle):

At this stage, you have gained some experience and are taking on more responsibilities. Your focus begins to shift to other things like:

- Building Networks: Establish relationships with candidates, clients, and other key industry stakeholders.
- Consulting on Client Needs: Understanding the client's requirements and providing guidance on how to meet their needs.
- Managing Full Cycle Recruitment: Overseeing the entire recruitment process, from sourcing to onboarding.

As an intermediate recruiter, you are developing your skills in areas like communication, negotiation, and problem-solving. You are working more closely with clients and candidates, and you are starting to build your professional network.

Advanced (Expert):

At this stage, you are now a seasoned recruiter with significant experience. Your responsibilities might include:

- Leading Teams: Managing and mentoring a team of recruiters, providing guidance and support.
- Contributing to Hiring Strategy: Providing input on the overall hiring strategy and helping to shape the recruitment process.
- Mentoring Others: Sharing your expertise and experience with junior recruiters, helping them develop their skills.
- Owning Client Relationships: Building and maintaining strong relationships with key clients, understanding their needs, and providing tailored solutions.

As an advanced recruiter, you're a trusted advisor to your clients and a leader in your organization. You're focused on delivering high-quality results, building strong relationships, and driving business growth.

Please note the following terms:

1. Progression: Each stage builds on the previous one, with increasing responsibilities and complexity.
2. Skill Development: As you progress, you are developing new skills, such as leadership, strategic thinking, and mentorship.
3. Relationship Building: Building strong relationships with clients, candidates, and colleagues is critical at every stage.
4. Adaptability: The recruitment industry is constantly evolving, so it is essential to stay adaptable and open to new challenges and opportunities.

CERTIFICATES AND COURSES THAT MATTER IN RECRUITMENT.

In the recruitment industry, having the proper certifications and courses can make a significant difference in your career. These credentials demonstrate your expertise and knowledge in specific areas of recruitment, making you a more attractive candidate to potential employers. Here are some of the most valuable certifications and courses that can benefit your career:

AIRS Certifications

The Advanced Internet Recruitment Seminars (AIRS) offers two certifications that are highly regarded in the industry:

1. Certified Internet Recruiter (CIR): This certification demonstrates your expertise in using the internet to recruit top talent. You'll learn how to effectively use online job boards, social media, and other digital tools to find and attract candidates.
2. Diversity Recruiter (DR): This certification focuses on promoting diversity and inclusion in the workplace. You will learn how to develop strategies to attract diverse candidates, reduce bias in the hiring process, and create a more inclusive work environment.

SHRM Talent Acquisition Specialty Credential

The Society for Human Resource Management (SHRM) offers a Talent Acquisition Specialty Credential that demonstrates your expertise in talent acquisition. This credential shows that you have the knowledge and skills to:

- Develop effective recruitment strategies
- Identify and attract top talent
- Conduct thorough interviews and assessments
- Make informed hiring decisions

LinkedIn Learning

LinkedIn Learning (formerly (Lynda.com) offers a wide range of courses on various topics, including:

- Recruiting: Learn how to develop effective recruitment strategies, use online job boards, and leverage social media to find top talent.
- Communication: Improve your communication skills to effectively engage with candidates, hiring managers, and other stakeholders.
- Employer Branding: Learn how to promote your employer brand and create a positive candidate experience.
- Diversity, Equity, and Inclusion (DEI): Understand the importance of DEI in the workplace and learn how to promote diversity and inclusion in your recruitment practices. (Reread chapter 15 to recall what DEI entails.)

HubSpot

HubSpot offers courses on inbound recruiting and recruitment marketing. These courses can help you learn how to:

- Attract top talent: Use inbound marketing strategies to attract potential candidates and promote your employer brand.
- Develop effective recruitment marketing campaigns: Learn how to create targeted recruitment marketing campaigns that drive results.
- Use data and analytics: Understand how to use data and analytics to measure the effectiveness of your recruitment marketing efforts.

Google Recruiter Certification

Google is developing a certification program for recruiters who use their recruitment tools. This certification will demonstrate your expertise in using Google's tools to recruit top talent. While the certification is still in development, it's expected to cover topics such as:

- Google's recruitment tools: Learn how to effectively use Google's recruitment tools, such as Google for Jobs and Google Cloud Talent Solution.
- Best practices for recruitment: Understand best practices for recruitment, including how to develop compelling job descriptions, conduct interviews, and make informed hiring decisions.

COMMUNITIES AND EVENTS FOR GROWTH.

As a recruiter, it is essential to stay connected with other professionals in the industry, learn from their experiences, and stay updated on the latest trends and practices. Communities and events can provide you with opportunities to network, learn, and grow in your career. Here are some communities and events that can benefit your career:

Online Forums

Online forums are an excellent way to connect with fellow recruiters, ask questions, and share knowledge. Some popular online forums for recruiters include:

- Recruiter.com: A community for recruiters to discuss various topics related to recruitment, including sourcing, interviewing, and hiring.

- SourceCon: A community for sourcers and recruiters to discuss tips, share knowledge, and network.
- HR communities on LinkedIn: LinkedIn groups dedicated to HR and recruitment, where professionals can share knowledge, ask questions, and network.

These online forums can provide you with:

- Access to a community of professionals: Connect with other recruiters, ask questions, and learn from their experiences.
- Knowledge sharing: Share your knowledge and expertise, and learn from others in the community.
- Networking opportunities: Expand your professional network and connect with potential employers, clients, or partners.

Virtual Webinars and Meetups

Virtual webinars and meetups are an excellent way to stay up-to-date on new trends and best practices in recruitment without leaving your office. Some popular virtual events for recruiters include:

- TalentNet: A platform that offers webinars and meetups on various topics related to recruitment, including talent acquisition, recruitment marketing, and diversity and inclusion.
- SHRM: The Society for Human Resource Management (SHRM) offers webinars and online events on various topics related to HR and recruitment.
- HCI: The Human Capital Institute (HCI) offers webinars and online events on various topics related to talent management and recruitment.

These virtual events can provide you with:

- Learning opportunities: Stay updated on the latest trends and best practices in recruitment, and learn from industry experts.
- Networking opportunities: Connect with other professionals in the industry, ask questions, and share knowledge.
- Convenience: Attend events from the comfort of your own office without having to travel.

Industry Events and Global Hiring Summits

Industry events and global hiring summits are excellent opportunities to network with other professionals, learn from industry experts, and stay current on the latest trends and technologies. Some popular industry events for recruiters include:

- In-person events: Attend conferences, summits, and workshops in person, where you can network with other professionals, learn from industry experts, and stay updated on the latest trends and technologies.
- Online events: Attend virtual conferences, summits, and workshops, where you can learn from industry experts, network with other professionals, and stay updated on the latest trends and technologies.

These industry events can provide you with the following:

- Networking opportunities: Connect with other professionals in the industry, potential employers, clients, or partners.
- Learning opportunities: Stay updated on the latest trends and best practices in recruitment, and learn from industry experts.
- Access to new technologies: Learn about new technologies and tools that can help you streamline your recruitment processes and improve your results.

By participating in online forums, virtual webinars and meetups, and industry events, you can stay connected with other professionals in the industry, learn from their experiences, and stay updated on the latest trends and best practices. This can help you advance in your career, enhance your skills, and achieve your objectives.

CREATING A PERSONAL GROWTH ROADMAP.

Creating a personal growth roadmap is a powerful way to take control of your career development and achieve your goals. It's like having a map that guides you through your journey, helping you stay focused and motivated. Here is a simple, step-by-step guide to creating your growth roadmap:

Step 1: Self-Assess

The first step in creating a personal growth roadmap is to assess your current skills and abilities. Ask yourself:

- What skills do I have? Make a list of your strengths and the skills you've developed over time. Consider both technical skills (like proficiency in recruitment software) and soft skills (like communication and problem-solving).
- What skills do I need?: Identify areas where you need improvement or would like to develop new skills. This could include skills specific to your industry or job function or more general skills such as leadership or time management.

By understanding your current strengths and weaknesses, you can create a personalized roadmap that aligns with your specific needs and goals.

Step 2: Set Short and Long-Term Goals

Once you have a clear understanding of your skills and abilities, it is now time to set some goals. Consider both short-term goals (what you want to achieve in the next 6-12 months) and long-term goals (what you want to accomplish in the next 1-5 years). Ask yourself:

- What do I want to achieve in my career? Do you want to move into a leadership role, specialize in a particular area of recruitment, or work internationally?
- What steps do I need to take to get there?: Break down your long-term goals into smaller, manageable steps. This will help you stay focused and motivated.

Some examples of short-term and long-term goals might include:

- Short-term goals:
  - Complete a certification program in recruitment
  - Take on additional responsibilities at work
  - Network with five new professionals in the industry
- Long-term goals:
  - Become a team leader or manager

- o Specialize in a particular area of recruitment (like diversity and inclusion or talent acquisition)
- o Work internationally or with global clients

Step 3: Invest Weekly

To achieve your goals, you need to invest time and effort each week. This might involve:

- Taking a course: Invest in your professional development by taking a course or attending a workshop. This could be online or in-person and may focus on specific skills or knowledge areas.
- Mentoring a junior recruiter: Share your knowledge and experience with others and help them develop their skills. This can be a great way to build your leadership skills and give back to the community.
- Writing a LinkedIn post: Share your thoughts and expertise with others on LinkedIn. This can help you build your professional brand and network with others in the industry.

The key is to commit to investing time each week in your professional development. Even small actions can add up over time and help you achieve your goals.

Step 4: Track Growth

Finally, it's essential to track your growth and progress over time. This will help you stay motivated and track your progress. Consider:

- Maintaining a journal: Write down your achievements, learnings, and reflections each week. This can help you process your experiences and identify areas for improvement.
- Using a digital record: Keep a digital record of your achievements, learnings, and reflections. This could be a spreadsheet, a note-taking app, or a digital journal.

By tracking your growth, you can:

- Celebrate your successes: Acknowledge and celebrate your achievements, no matter how small they may seem.
- Identify areas for improvement: Reflect on your experiences and identify areas where you need to improve or develop new skills.

- Adjust your roadmap: Use your reflections to refine your roadmap and make necessary adjustments. This will help you stay on track and achieve your goals.

By following these steps, you can create a personal growth roadmap that helps you achieve your goals and advance your career. Remember to stay flexible, be patient, and celebrate your successes along the way!

HABITS THAT ACCELERATE PROFESSIONAL GROWTH.

Developing good habits can significantly impact your professional growth and success. By incorporating certain habits into your daily routine, you can accelerate your learning, build your confidence, and achieve your goals. Here are four habits that can help you grow professionally:

1. Block Time Every Week for Personal Learning

One of the most essential habits for professional growth is dedicating time to personal learning. This can involve:

- Reading industry publications: Stay up to date on the latest trends and practices in your industry by reading relevant journals, blogs, and books.
- Taking online courses: Invest in your professional development by taking online courses or attending webinars that focus on specific skills or knowledge areas.
- Watching tutorials and videos: Watch tutorials, videos, or podcasts that provide insights and tips on topics related to your profession.

By blocking time every week for personal learning, you can:

- Stay current with industry developments: Stay ahead of the curve by learning about the latest trends, technologies, and best practices in your industry.
- Develop new skills: Acquire new skills and knowledge that can help you perform your job more effectively and take on new challenges.
- Enhance your professional credibility: Demonstrate your commitment to professional development and enhance your credibility with colleagues, clients, and peers.

2. Reflect on Both Wins and Mistakes

Reflection is a powerful tool for professional growth. By reflecting on both wins and mistakes, you can:

- Identify areas for improvement: Analyze your mistakes and identify areas where you need to improve or develop new skills.
- Build on successes: Reflect on your wins and identify what worked well so you can build on those successes in the future.
- Develop a growth mindset: Cultivate a growth mindset by embracing challenges, learning from failures, and persisting in the face of obstacles.

To reflect effectively, try:

- Keeping a journal: Write down your thoughts, feelings, and insights after each project or significant event.
- Conducting regular self-assessments: Take time to reflect on your progress, goals, and areas for improvement.
- Seeking feedback: Ask for feedback from others to gain new perspectives and insights.

3. Ask for Feedback from Peers, Candidates, and Clients

Feedback is essential for professional growth. By asking for feedback from peers, candidates, and clients, you can:

- Gain new perspectives: Get insights and perspectives from others that can help you identify areas for improvement and develop new skills.
- Improve your performance: Use feedback to adjust your approach, strategies, and tactics to achieve better results.
- Build stronger relationships: Demonstrate that you value others' opinions and are committed to growth and improvement, which can foster stronger relationships.

To ask for feedback effectively, try:

- Being specific: Ask specific questions that can help you gain actionable insights.
- Being open-minded: Listen to feedback without becoming defensive or dismissive.

- Following up: Follow up on feedback by implementing changes or adjustments.

4. Share Knowledge to Build Your Brand and Confidence

Sharing knowledge is a powerful way to build your professional brand and confidence. By sharing your expertise and insights, you can:

- Establish yourself as a thought leader: Demonstrate your expertise and establish yourself as a thought leader in your industry.
- Build your professional network: Connect with others who share similar interests and passions.
- Develop your communication skills: Improve your ability to articulate your ideas and communicate effectively.

To share knowledge effectively, try:

- Writing articles or blog posts: Share your insights and expertise through written content.
- Speaking at events or webinars: Share your knowledge and expertise through public speaking.
- Participating in online communities: Engage with others in online communities related to your profession.

By incorporating these habits into your daily routine, you can accelerate your professional growth, build your confidence, and achieve your goals.

CASE EXAMPLE: A recruiter in Oceania started journaling every Friday about her top 3 wins and 1 lesson. In one year, she had a portfolio of stories that helped her land a promotion to lead recruiter and become a speaker at an industry webinar.

NOTE: This example is scenario-based training and learning, which helps us understand the scope better. The scope can be much deeper and more precise.

Self Assessment

1. What are two signs that you're ready to move into a leadership role?
2. Name one certification that can enhance a recruiter's career.
3. Why is reflection an essential habit for career growth?

4. What's one way to build your professional brand?
5. How often should you invest time in learning something new?

Reflective Prompt: Think about where you want to be in your recruitment career one year from now. What three actions can you start this month to get closer to that vision?

CHAPTER NINETEEN

FUTURE OF GLOBAL TALENT ACQUISITION

Objectives of this Chapter:

- Understand key trends shaping the future of IT recruitment
- Explore the impact of remote work and virtual hiring models
- Learn how to navigate cross-cultural collaboration in global teams
- Examine the evolving role of AI in partnership with human recruiters
- Prepare for changes in candidate behavior, platforms, and client needs

Talking about the future of recruitment. The truth is, the future starts now. The global recruitment industry has undergone a significant transformation over the past five years, more than it did in the previous twenty years.

We've seen a shift towards fully remote tech teams, where people can work from anywhere in the world. We've also seen the rise of AI-powered matching engines that can help us quickly and efficiently find the best talent.

The future of talent acquisition is flexible and digital, with boundariesboundaries, meaning limitedlimited. It's about being adaptable, using technology to our advantage, and being open to new ways of working.

So, what does this mean for you? It means that you need to be prepared to stay ahead of the curve. You need to be willing to learn, adapt, and innovate.

This chapter will help you do just that. It will prepare you to thrive in a rapidly changing industry where flexibility, digital skills, and a global perspective are essential.

The future is here, and it's incredibly exciting. Now, let's seize the opportunities it presents and shape the future of recruitment together.

REMOTE HIRING: Here to Stay.

Remote hiring has become an integral part of the recruitment industry, and it is here to stay. With the advancement of technology and the shift towards remote work, virtual interviews, onboarding, and remote assessments have become the norm. We will discuss what this means and how it is currently impacting the recruitment industry.

1. Virtual Interviews

Virtual interviews are interviews that take place remotely, often via video conferencing software, such as Google MeetMeet, Zoom, Skype, and others. This allows recruiters to conduct interviews with candidates from anywhere in the world at any time. Virtual interviews have several benefits, including:

- Increased flexibility: Virtual interviews can be scheduled at any time, making it easier for both recruiters and candidates to find a time that works for them.
- Reduced costs: Virtual interviews eliminate the need for travel, reducing costs for both the recruiter and the candidate.
- Broader reach: Virtual interviews allow recruiters to interview candidates from anywhere in the world, increasing the pool of potential candidates.

2. Onboarding

Onboarding is the process of integrating new employees into a company. With remote hiring, onboarding has also become a virtual process. This includes:

- Virtual orientation: New employees can participate in virtual orientation sessions, where they are introduced to the company culture, policies, and procedures.
- Digital documentation: New employees can complete digital paperwork and documentation, reducing the need for physical paperwork.
- Virtual training: New employees can participate in virtual training sessions, where they learn about their roles, responsibilities, and expectations.

3. Remote Assessments

Remote assessments are used to evaluate a candidate's skills and abilities remotely. This can include:

- Online skills tests: Candidates can complete online skills tests to demonstrate their abilities.

- Virtual simulations: Candidates can participate in virtual simulations that mimic real-world scenarios, allowing recruiters to assess their skills practically.
- Video interviews: Video interviews can be used to assess a candidate's communication skills, personality, and cultural fit.

4. Global Recruitment Opportunities.

Remote hiring has opened up opportunities for global recruiters. With virtual interviews, onboarding, and remote assessments, recruiters can now work with clients and candidates from anywhere in the world. This means:

- Access to global talent: Recruiters can tap into an international pool of talent, thereby increasing the likelihood of finding the ideal candidate for the job.
- Increased opportunities: Remote hiring has expanded the possibilities for recruiters to collaborate with clients and candidates from diverse regions and cultures.
- Cultural diversity: Remote hiring has introduced cultural diversity to the recruitment industry, enabling recruiters to collaborate with individuals from diverse backgrounds and perspectives.

5. Flexible Work from Anywhere Setups

The shift towards remote hiring has also led to more flexible work-from-anywhere setups. This means:

- Work from anywhere: Recruiters can work from anywhere, at any time, as long as they have a stable internet connection.
- Flexible schedules: Recruiters can create their schedules, allowing them to balance work and personal life more effectively.
- Asynchronous workflows: Recruiters can work asynchronously, meaning they can complete tasks and respond to messages at their own pace rather than in real-time.

6. Asynchronous Workflows

Asynchronous workflows are workflows that don't require real-time communication. This means that it has:

- Reduced meetings: Recruiters can reduce the number of sessions they attend, allowing them to focus on more critical tasks.
- Increased productivity: Recruiters can work more efficiently, completing tasks and responding to messages at their own pace.
- Improved work-life balance: Asynchronous workflows allow recruiters to balance work and personal life more effectively, reducing the need for overtime and improving overall well-being.

In conclusion, remote hiring is here to stay and is undergoing significant transformation in the recruitment industry. With virtual interviews, onboarding, and remote assessments, recruiters can now work with clients and candidates from anywhere in the world. The shift towards remote hiring has also led to more flexible work-from-anywhere setups and asynchronous workflows, allowing recruiters to work more efficiently and effectively.

CROSS-CULTURAL COLLABORATION SKILLS.

In today's globalized world, recruiters must possess cross-cultural collaboration skills to work effectively with clients and candidates from diverse cultural backgrounds. This means understanding the cultural nuances that can impact communication, work ethic, and time expectations. Let's break down what this means and how recruiters can develop these skills.

1. Bridging Cultural Gaps

When working with clients and candidates from diverse cultural backgrounds, recruiters must be aware of the potential cultural differences that can impact communication and collaboration. These gaps can include:

- Tone: Different cultures have distinct communication styles, and recruiters must be aware of these differences to prevent misunderstandings.
- Work ethic: Different cultures have varying work ethics, and recruiters must understand these differences to ensure that candidates are a good cultural fit for the company.
- Time expectations: Different cultures have varying expectations regarding time and punctuality, and recruiters must be aware of these differences to avoid misunderstandings.

2. Understanding Regional Collaboration Styles

Understanding how teams in different regions collaborate can help recruiters screen and prepare candidates more effectively. For example:

- Communication styles: In some cultures, direct and assertive communication is valued, while in others, more indirect and polite communication is preferred.
- Teamwork: In some cultures, teamwork is highly valued, while in others, individual achievement is more important.
- Decision-making: In some cultures, decision-making is a collaborative process, while in others, decisions are made more quickly and individually.

Tools for Cross-Cultural Collaboration

There are several tools that recruiters can use to develop their cross-cultural collaboration skills, including:

- Hofstede's Cultural Dimensions: This framework offers a means to understand cultural differences across various regions and countries.
- Slack etiquette guides: These guides offer tips and best practices for effective communication on Slack, a popular collaboration tool.
- Virtual onboarding resources: These resources offer guidance on how to onboard new employees remotely, including tips for effective communication and setting clear expectations.

Benefits of Cross-Cultural Collaboration Skills:

Developing cross-cultural collaboration skills can have several benefits for recruiters, including:

- Improved communication: By understanding cultural nuances, recruiters can communicate more effectively with clients and candidates from diverse cultural backgrounds.
- Better candidate fit: By understanding cultural differences, recruiters can more effectively match candidates with company cultures, thereby reducing turnover and enhancing job satisfaction.
- Increased client satisfaction: By understanding cultural differences, recruiters can better meet the needs of clients from diverse cultural backgrounds, increasing client satisfaction and loyalty.

Developing Cross-Cultural Collaboration Skills:

Recruiters can develop cross-cultural collaboration skills by:

- Learning about different cultures: Recruiters can learn about other cultures by reading books, articles, and online resources.
- Practicing cross-cultural communication: Recruiters can develop their cross-cultural communication skills by working with clients and candidates from diverse cultural backgrounds.
- Using cultural intelligence tools: Recruiters can utilize tools like Hofstede's Cultural Dimensions to understand cultural differences better and develop their cross-cultural collaboration skills.

In conclusion, cross-cultural collaboration skills are crucial for recruiters operating in today's increasingly globalized world. By understanding cultural nuances and developing these skills, recruiters can improve communication, better match candidates with company cultures, and increase client satisfaction.

HUMAN + AI COLLABORATIONS.

The recruitment industry is undergoing a significant transformation with the integration of Artificial Intelligence (AI) and human recruiters. While AI excels at automating repetitive and data-intensive tasks, human recruiters bring essential skills, such as empathy, context, and storytelling, to the table. So, how can human-AI collaborations revolutionize the recruitment process? We will talk about that now.

AI's Role in Recruitment:

AI can automate various tasks in the recruitment process, including:

- Screening: AI can quickly sift through resumes and cover letters to identify top candidates based on keywords, experience, and qualifications.
- Scheduling: AI-powered scheduling tools can automatically schedule interviews and send reminders to candidates and interviewers.
- Initial engagement: AI-powered chatbots can engage with candidates, answer their questions, and provide them with information about the company and the role.

Human Recruiters' Role:

While AI excels at automating tasks, human recruiters bring essential skills that AI currently can't replicate. These skills include:

- Empathy: Human recruiters can empathize with candidates, understand their concerns, and provide personalized support.
- Context: Human recruiters can provide valuable context to the recruitment process, including an understanding of the company culture, job requirements, and team dynamics.
- Storytelling: Human recruiters can tell stories about the company, the role, and the team, making the job more appealing to candidates.

Benefits of Human + AI Collaborations

The collaboration between humans and AI can bring numerous benefits to the recruitment process, including:

- Increased efficiency: AI can automate repetitive tasks, freeing up human recruiters to focus on high-touch, high-value tasks.
- Improved candidate experience: AI can provide candidates with quick and efficient communication, while human recruiters can provide personalized support and empathy.
- Better hiring decisions: Human recruiters can use AI-generated insights to make more informed hiring decisions, combining data-driven insights with human judgment.

Upskilling in AI Tools.

To succeed in the new era of human-AI collaborations, recruiters need to upskill in AI tools and learn how to work effectively with AI. This includes:

Understanding AI-powered tools: Recruiters must comprehend how AI-powered tools function, including their strengths and limitations.

- Using AI-generated insights: Recruiters need to learn how to use AI-generated insights to inform their hiring decisions.
- Developing human skills: Recruiters need to develop their human skills, including empathy, context, and storytelling, to complement the capabilities of AI.

Reinforcing Human Judgment:

While AI excels at analyzing data, human judgment remains essential for making nuanced hiring decisions. Recruiters need to:

- Use AI-generated insights judiciously: Recruiters should use AI-generated insights as a guide, rather than relying solely on AI to make hiring decisions.
- Consider the context: Recruiters must consider the job's context, the company's culture, and the team's dynamics when making hiring decisions.
- Trust their instincts: Recruiters need to trust their instincts and use their human judgment to make hiring decisions that are in the best interest of the company.

In conclusion, human-AI collaborations have the potential to revolutionize the recruitment process. By automating repetitive tasks and providing data-driven insights, AI can free up human recruiters to focus on high-touch, high-value tasks. By upskilling in AI tools and reinforcing human judgment, recruiters can thrive in this new era of human-AI collaborations.

EMERGING TRENDS TO WATCH.

The recruitment industry is constantly evolving, and several emerging trends are worth watching. These trends have the potential to disrupt traditional staffing models, improve hiring practices, and create more inclusive and diverse workplaces. Let's explore these trends in more detail.

1. Talent Marketplaces

Talent marketplaces are platforms that connect freelance talent with businesses that need specific skills. Examples of talent marketplaces include Toptal, Upwork Pro, and Braintrust. These platforms are disrupting traditional staffing models by providing businesses with access to a global pool of talent.

- Benefits: Talent marketplaces offer businesses flexibility, cost savings, and access to specialized skills.
- Challenges: Talent marketplaces can make it difficult for businesses to find the right talent and may require new skills and strategies for managing remote workers.

2. Blockchain for Verifiable Work Histories

Blockchain technology is being used to create verifiable work histories for employees. This means that employees can have a secure and tamper-proof record of their work experience, skills, and achievements.

- Benefits: Blockchain-based work histories can reduce the risk of CV fraud, improve the accuracy of background checks, and provide employees with more control over their data.
- Challenges: The adoption of blockchain technology in recruitment is still in its early stages, and it may face regulatory and technical challenges.

3. Neurodiversity and Accessibility in Hiring

There is a growing focus on neurodiversity and accessibility in hiring, with many companies recognizing the benefits of creating a diverse and inclusive workplace.

- Benefits: Neurodiverse teams can bring unique perspectives and skills to the workplace, thereby enhancing innovation and creativity.
- Challenges: Companies must create an inclusive culture and provide accommodations for neurodiverse employees, which can necessitate significant investment and effort.

4. Fractional Hiring

Fractional hiring involves hiring professionals on a part-time or project basis rather than full-time. This can include hiring part-time CTOs or project-based specialists.

- Benefits: Fractional hiring can provide businesses with flexibility, cost savings, and access to specialized skills.
- Challenges: Fractional hiring can require new skills and strategies for managing remote workers and may require businesses to adapt their workflows and processes.

5. DEI becoming part of Every Hiring Metric.

Diversity, equity, and inclusion (DEI) are becoming increasingly important in hiring, with many companies recognizing the benefits of creating a diverse and inclusive workplace.

- Benefits: DEI initiatives can enhance employee engagement, retention, and productivity while also positively impacting a company's reputation and brand.
- Challenges: Implementing effective DEI initiatives can require significant investment and effort and may require businesses to challenge their existing biases and practices.

In conclusion, these emerging trends have the potential to transform the recruitment industry, creating more inclusive and diverse workplaces. By understanding these trends and adapting to the changing landscape, businesses can stay ahead of the curve and attract the best talent.

HOW TO STAY AHEAD.

In today's fast-paced recruitment industry, it's essential to stay ahead of the curve to remain competitive. Here are some strategies to help you stay ahead:

1. Follow Global HR and Recruiting Forums

Following global HR and recruiting forums is an excellent way to stay informed about the latest trends, best practices, and industry developments. Some popular forums include:

- SHRM (Society for Human Resource Management): SHRM is a professional organization for HR professionals that provides resources, training, and networking opportunities.
- SourceCon: SourceCon is a recruitment conference and community that focuses on sourcing and talent acquisition.
- RecruitingDaily: RecruitingDaily is a recruitment blog and newsletter that provides news, insights, and best practices for recruiters.

By following these forums, you can:

- Stay up to date on industry trends*: Learn about the latest developments and trends in recruitment and HR.
- Network with professionals: Connect with other professionals in the industry and learn from their experiences.

Enhance your skills by Utilizing training and resources to refine your knowledge and expertise.

2. Subscribe to AI, HRTech, and Talent Strategy Newsletters

Subscribing to newsletters is an excellent way to stay informed about the latest developments in AI, HRTech, and talent strategy. Some popular newsletters include:

- AI and HRTech newsletters: These newsletters provide insights and updates on the latest AI and HRTech trends and innovations.
- Talent strategy newsletters: These newsletters provide insights and best practices on talent acquisition, management, and development.

By subscribing to these newsletters, you can:

- Stay informed about the latest trends: Learn about the latest developments and trends in AI, HRTech, and talent strategy.
- Get insights from experts: Read insights and opinions from industry experts and thought leaders.
- Enhance your knowledge: Expand your understanding of AI, HR Tech, and talent strategy.

3. Keep Experimenting with Tools, Formats, and Outreach Methods

The recruitment industry is constantly evolving, and it's essential to stay ahead by experimenting with new tools, formats, and outreach methods. This includes:

- Trying new recruitment tools: Experiment with new recruitment tools and technologies to improve efficiency and effectiveness.
- Testing different formats: Test different formats, such as video interviews or gamification, to improve candidate engagement and experience.
- Exploring new outreach methods: Explore new outreach methods, such as social media or employee referrals, to attract top talent.

By experimenting with new tools, formats, and outreach methods, you can:

- Improve efficiency: Streamline processes and improve efficiency by leveraging new tools and technologies.

Enhance candidate experience: Improve the candidate experience by utilizing innovative formats and outreach methods.

- Stay competitive: Leverage the latest trends and innovations in recruitment to stay ahead of the curve.

4. Focus on Adaptability, Emotional Intelligence, and Data Fluency

To succeed in the recruitment industry, it is essential to focus on adaptability, emotional intelligence, and data fluency. This includes:

- Adaptability: Be open to change and willing to adapt to new trends, technologies, and best practices.
- Emotional intelligence: Develop emotional intelligence to understand better and connect with candidates, clients, and colleagues.
- Data fluency: Develop data fluency to analyze and interpret data and make informed decisions.

By focusing on adaptability, emotional intelligence, and data fluency, you can:

- Improve performance: Enhance performance by being adaptable, empathetic, and data-driven.
- Build strong relationships: Build strong relationships with candidates, clients, and colleagues by being empathetic and understanding.
- Make informed decisions: Make informed decisions by analyzing and interpreting data.

In conclusion, staying ahead in the recruitment industry requires a commitment to ongoing learning, experimentation, and innovation. By following global HR and recruiting forums, subscribing to AI, HRTech, and talent strategy newsletters, experimenting with tools, formats, and outreach methods, and focusing on adaptability, emotional intelligence, and data fluency, you can stay competitive and achieve success in the industry.

CASE EXAMPLE: A recruiter in Africa shifted their model entirely to remote sourcing and pre-screening. They might have done that because he wasn't getting the kind of Candidates he wanted within his vicinity. He had to shift his model to remote so he could get applications from three other countries. This helped clients from other countries fill remote DevOps roles with candidates from different parts of the world. Their success increased after they began using Slack, Google Meet, and asynchronous updates and when they started working within their time zones to make things more convenient for everyone.

NOTE: This example is scenario-based for training and learning purposes to help us understand the scope better. The real scope can be much deeper and more precise.

Self Assessment

1. Name two major trends shaping the future of global recruitment.
2. What is one challenge of cross-cultural collaboration?
3. Why is it essential to upskill in AI while reinforcing the human touch?
4. What is fractional hiring, and why is it gaining popularity?
5. How can recruiters stay informed about upcoming industry shifts?

Reflective Prompt: Where do you see your recruitment career going in the next 3-5 years? What future trends are you most excited or concerned about, and why?

CHAPTER TWENTY

REAL RECRUITERS, REAL STORIES

Objectives of this Chapter:

- Learn from real experiences of global recruiters placing talent in the US IT industry
- Identify success factors that transcend borders and individual backgrounds
- Recognize the everyday struggles and breakthroughs recruiters face worldwide
- Discover practical insights, tips, and habits from recruiters at different career stages
- Reflect on your approach by comparing it with those who've walked a similar path

Why does this Chapter matter? It is about people's real-life experiences, and there is a saying that "Experience is the Best teacher". When people talk to you about their real-life experiences, it is more believable and relatable. It makes you feel it and understand the raw feeling of what a person has gone through. It makes you feel sad during their difficult moments, it motivates you through the ups and downs, and it fills you with joy when they finally succeed. It also conveys their anger and hurt, and it makes c, concepts more realistic. When you hear about something that has happened, of course, you are going to believe it easier.

In this Chapter, we will discuss the various stories of individuals and recruiters from different continents around the world. Being a recruiter is different on each continent; how it works in Asia isn't the same as it is in America, and this is what this Chapter is all about. Recruiters will discuss their challenges, how they were resolved, and the lessons they learned from their experiences.

VOICES FROM RECRUITERS ACROSS ASIA, EUROPE, OCEANIA AND AMERICA.

Asia (Southern Region):

A recruiter was sourcing for IT project managers with no background in tech. She knew nothing about tech, but she wanted to be a recruiter. She used mock interviews, shadowing peers and whiteboard mapping to build on her knowledge. She kept doing this until she had learnt a lot about tech in IT. Today, she leads a regional talent team which serves approximately 30 clients based in the US alone.

Africa (Eastern Region):

A recruiter had no resources or money, but that did not stop him from building an Excel tracker that mimics an Applicant Tracking System; even with his lack of resources, he was able to pull off that feat, and within six weeks, he was able to place a cloud engineer to be in charge of it. That recruiter now consuls on sourcing strategies for other companies and firms because of the name he has created for himself.

Europe (Western Region): A former Human Resource assistant who switched to a recruiter who uses multilingual skills to assist European candidates in understanding the US hiring system whenever they need assistance. She helped reduce the language barrier in Europe, and now she is the top biller at her agency.

Oceania (Pacific Region):

A remote recruiter who had worked for various global brands had the idea of cutting interviews, and he brought that idea to life by creating video screening libraries, which he used to cut clients' interviews in half. That was a very brilliant innovation, and it led to him bagging a partnership model with one of the largest software firms in North America.

Americas (South and Central):

A recruiter worked in a rural town with no civilization; he searched for overlooked tech talents using job boards and searching through social media; after identifying these talents, he built a niche agency with a 95% interview-to-offer ratio, which means that you have a 95% chance of getting the offer when you interview with him.

COMMON SUCCESS PATTERNS FROM GLOBAL RECRUITERS.

Global recruiters have developed various strategies and habits that contribute to their success in finding and placing top talent. Let's explore some common success patterns that can help recruiters improve their skills and achieve their goals.

1. Daily Sourcing Hours

Many successful recruiters dedicate specific hours each day to sourcing candidates. This involves:

- Searching for candidates: Using various tools and platforms, such as LinkedIn, job boards, and social media, to find potential candidates.
- Reaching out to candidates: Contacting candidates directly to gauge their interest in a role and assess their qualifications.
- Building a pipeline: Creating a pool of potential candidates for future job openings.

By dedicating specific hours to sourcing, recruiters can:

- Stay consistent: Consistency is key in sourcing, and dedicating specific hours each day helps recruiters stay on track.
- Improve efficiency: By focusing on sourcing during specific hours, recruiters can improve their efficiency and productivity.
- Find better candidates: Regular sourcing efforts can help recruiters identify and attract top candidates, thereby increasing their chances of making successful placements.

2. Use of Templates and Trackers

Successful recruiters often utilize templates and trackers to streamline their workflow and enhance productivity. This includes:

- Email templates: Using pre-written email templates to communicate with candidates and clients.
- Candidate trackers: Using spreadsheets or other tools to track candidate progress and status.

- Interview guides: Using guides to structure interviews and ensure that all necessary topics are covered.

By using templates and trackers, recruiters can:

- Save time: Templates and trackers can save recruiters time and effort, allowing them to focus on high-value tasks.
- Improve consistency: Templates and trackers can help recruiters maintain consistency in their communication and workflow.
- Reduce errors: By using templates and trackers, recruiters can reduce errors and ensure that all necessary steps are taken.

3. Intake Call Discipline

Intake calls are an essential part of the recruitment process, and successful recruiters approach these calls with discipline. This includes:

- Preparing for the call: Research the client and the job requirements to ensure a thorough understanding of the role.
- Asking the right questions: Asking questions to clarify the client's needs and expectations.
- Taking notes: Taking detailed notes during the call to ensure that all necessary information is captured.

By approaching intake calls with discipline, recruiters can:

- Understand client needs: Gain a deep understanding of the client's needs and expectations.

Set clear expectations: Communicate the recruitment process and timeline to the client.

- Build trust: Build trust with the client by demonstrating professionalism and expertise.

4. Continuous Upskilling

The recruitment industry is constantly evolving, and successful recruiters recognize the importance of ongoing professional development and continuous skill enhancement. This includes:

- Watching webinars: Staying up to date with industry trends and best practices by watching webinars and online training sessions.
- Reading job specs: Staying informed about job requirements and industry developments by reading job specs and industry publications.
- Learning new skills: Acquiring new skills and knowledge to improve performance and stay competitive.

By continuously upskilling, recruiters can:

- Stay current: Stay current with industry trends and best practices.
- Improve performance: Improve performance by acquiring new skills and knowledge.
- Stay competitive: Stay competitive in a rapidly changing industry.

5. Soft Skill Development

In addition to technical skills, successful recruiters also develop essential soft skills that contribute to their success. This includes:

- Active listening: Listening carefully to clients and candidates to understand their needs and concerns.
- Email writing: Writing effective emails that are clear, concise, and professional.
- Candidate prep: Preparing candidates for interviews and ensuring they have the information they need to succeed.

By developing soft skills, recruiters can:

- Build strong relationships: Build strong relationships with clients and candidates by demonstrating empathy and understanding.
- Communicate effectively: Communicate clearly and concisely with clients and candidates, minimizing misunderstandings and errors.
- Provide excellent service: Provide exceptional service to clients and candidates, leading to increased satisfaction and loyalty.

In conclusion, successful recruiters have developed various strategies and habits that contribute to their success. By adopting these common success patterns, recruiters can enhance their skills, boost productivity, and achieve their objectives.

"PASSION AND CURIOSITY MATTER MORE THAN A TECHNICAL DEGREE."

When it comes to succeeding in the recruitment industry, many people believe that having a technical degree is essential. However, the truth is that passion and curiosity can be just as important, if not more so. Let's explore why this is the case.

What Do We Mean by Passion and Curiosity?

Passion and curiosity refer to a person's enthusiasm and interest in learning and exploring new things. In the context of recruitment, this means being passionate about helping people find jobs, understanding what motivates them, and being curious about different industries and roles.

Why Are Passion and Curiosity Important?

Passion and curiosity are essential in recruitment because they drive recruiters to:

- Stay up to date with industry trends: Recruiters who are passionate about their work and curious about industry trends are more likely to stay informed about the latest developments and best practices.
- Build strong relationships: Passionate and curious recruiters are more likely to build strong relationships with clients and candidates, understanding their needs and providing personalized support.
- Think creatively: Passion and curiosity can help recruiters think creatively and devise innovative solutions to complex problems.

How Can Passion and Curiosity Be Developed?

While some people may naturally be more passionate and curious than others, these traits can also be developed over time. Here are some ways to cultivate passion and curiosity:

- Explore different industries and roles: Learning about different industries and roles can help recruiters develop a deeper understanding of the job market and stay curious about new developments.
- Ask questions: Asking questions and seeking feedback can help recruiters stay curious and improve their skills.
- Stay up to date with industry trends: Reading industry publications, attending conferences, and participating in online forums can help recruiters stay informed and passionate about their work.

The Limitations of a Technical Degree:

While a technical degree can provide a solid foundation in a particular field, it may not be enough to guarantee success in recruitment. Here are some limitations of relying solely on a technical degree:

- Limited transferable skills: Technical skills may not be directly transferable to recruitment, and recruiters may need to develop additional skills, such as communication and interpersonal skills, to succeed in their roles.
- Industry changes rapidly: The recruitment industry is constantly evolving, and technical skills may become outdated quickly.
- Soft skills are essential: Recruitment requires strong soft skills, such as communication, empathy, and problem-solving, which may not be fully developed through a technical degree.

The Benefits of Passion and Curiosity

Passion and curiosity can bring numerous benefits to recruiters, including:

- Increased job satisfaction: Recruiters who are passionate about their work are more likely to be satisfied with their job and stay motivated.
- Improved performance: Passion and curiosity can drive recruiters to perform at a higher level, staying up to date with industry trends and best practices.
- Better relationships: Passionate and curious recruiters are more likely to build strong relationships with clients and candidates, as they understand their needs and provide personalized support.

In conclusion, while a technical degree can provide a solid foundation, passion and curiosity are essential for success in recruitment. By cultivating these traits, recruiters can stay up-to-date with industry trends, build strong relationships, and think creatively.

SYSTEM CREATES CONSISTENCY (Even simple ones)

In recruitment, consistency is key to achieving success. One way to ensure consistency is by implementing simple systems. Let's explore how systems can create consistency and improve recruitment processes.

What Are Systems?

Systems refer to a set of processes or procedures that are followed consistently to achieve a specific goal. In recruitment, systems can include:

- Sourcing processes: A systematic approach to finding and contacting potential candidates.
- Interview processes: A standardized process for conducting interviews, including questions and evaluation criteria.
- Communication protocols: A consistent approach to communicating with candidates and clients.

How Do Systems Create Consistency?

Systems create consistency by:

- Standardizing processes: Systems ensure that processes are followed consistently, reducing variability and errors.
- Reducing reliance on individual memory: Systems document processes and procedures, reducing dependence on personal memory and minimizing the risk of mistakes.
- Improving efficiency: Systems streamline processes, making them more efficient and reducing the time required to complete tasks.

Benefits of Systems in Recruitment

Implementing systems in recruitment can bring numerous benefits, including:

- Improved candidate experience: Consistent processes ensure that candidates receive a similar experience, regardless of the recruiter or role.
- Increased efficiency: Systems reduce the time required to complete tasks, allowing recruiters to focus on high-value activities.
- Better decision making: Systems provide a framework for making decisions, reducing the risk of bias and errors.
- Scalability: Systems enable recruitment teams to scale their processes, handling increased volume without sacrificing quality.

Examples of Simple Systems

Simple systems can be just as effective as complex ones. Here are some examples:

- Using a template for candidate emails: A template ensures that all candidates receive consistent communication, reducing errors and improving efficiency.

Creating a checklist for interviews ensures that all necessary topics are covered during the interview, reducing the risk of mistakes.

Implementing a standardized evaluation process ensures that all candidates are assessed consistently, thereby reducing bias and errors.

Implementing Systems

Implementing systems requires:

- Identifying areas for improvement: Identifying areas where consistency is lacking and determining the root cause of the issue.
- Designing a system: Designing a system that addresses the issue and improves consistency.
- Training and documentation: Training recruiters on the new system and documenting processes to ensure consistency.
- Monitoring and evaluation: Continuously monitoring and evaluating the system to ensure it is working effectively.

Best Practices for Implementing Systems

Best practices for implementing systems include:

- Keep it simple: Start with simple systems and gradually build complexity.
- Involve stakeholders: Involve recruiters and other stakeholders in the design and implementation of systems.
- Continuously evaluate and improve: Regularly evaluate and improve systems to ensure they remain effective.

In conclusion, systems create consistency in recruitment by standardizing processes, reducing reliance on individual memory, and improving efficiency. By implementing simple systems, recruiters can enhance the candidate experience, increase efficiency, and make more informed decisions.

SUCCESS OFTEN COMES AFTER 3-6 MONTHS OF DISCIPLINED TRIAL AND ERROR.

When it comes to achieving success in recruitment, many people believe that it's a quick fix or an overnight success. However, the reality is that success often comes after a period of disciplined trial and error. Let's explore what this means and how it can help recruiters achieve their goals.

What is Disciplined Trial and Error?

Disciplined trial and error refers to a systematic approach to testing and refining different strategies and techniques. In recruitment, this might involve:

- Testing different sourcing channels: Trying out different sourcing channels, such as social media or job boards, to see which ones work best.
- Experimenting with different interview questions: Testing different interview questions to see which ones are most effective in assessing candidate fit.
- Refining communication scripts: Refining communication scripts to ensure that they are clear, concise, and practical.

Why is Disciplined Trial and Error Important?

Disciplined trial and error is necessary because it allows recruiters to:

- Learn from mistakes: By trying new approaches and testing different strategies, recruiters can learn from their mistakes and refine their approach.
- Identify what works: Disciplined trial and error helps recruiters identify what works and what doesn't, allowing them to focus on the most effective strategies.
- Improve efficiency: By refining processes and strategies, recruiters can improve efficiency and reduce the time required to complete tasks.

The 3-6 Month Timeline

The 3-6 month timeline refers to the period it often takes to see significant results from disciplined trial and error. This timeline can vary depending on individual circumstances, but it provides a general framework for understanding the process.

- Months 1-3: Experimentation: During the first few months, recruiters experiment with different strategies and techniques, testing what works and what doesn't.
- Month 4-6: Refining and Optimizing: As recruiters gain more experience and data, they can refine and optimize their approach, focusing on the most effective strategies and techniques.

Benefits of Disciplined Trial and Error

The benefits of disciplined trial and error include:

- Improved results: By testing and refining various strategies, recruiters can enhance their outcomes and achieve their objectives more effectively.
- Increased confidence: Disciplined trial and error helps recruiters build confidence in their abilities and develop a sense of mastery.
- Adaptability: By being open to trying new approaches and testing different strategies, recruiters can adapt to changing circumstances and stay ahead of the curve.

Best Practices for Disciplined Trial and Error

Best practices for disciplined trial and error include:

- Set clear goals: Setting clear goals and objectives helps recruiters stay focused and motivated.
- Track progress: Tracking progress and results helps recruiters identify what's working and what areas need improvement.
- Stay flexible: Being open to trying new approaches and testing different strategies helps recruiters remain adaptable and respond to changing circumstances.

In conclusion, success in recruitment often comes after a period of disciplined trial and error. By testing and refining various strategies and techniques, recruiters can enhance their results, boost their confidence, and adapt to shifting circumstances. By following best practices and staying committed to the process, recruiters can achieve their goals and succeed in the industry.

RELATIONSHIP BUILDING WITH CANDIDATES.

In the world of recruitment, it's easy to get caught up in the numbers game, which is finding the perfect candidate with the ideal resume. However, relationship-building with candidates is just as valuable as resume matching. Let's explore why.

What is Relationship-Building?

Relationship building is the process of establishing trust, rapport, and communication with candidates. It's about creating a positive experience for them, from the initial contact to the final decision. This approach helps build a strong connection between the candidate, the recruiter, and the company.

Why is Relationship Building Important?

1. Candidate Experience: A positive relationship with the recruiter can significantly impact a candidate's overall experience. Candidates who feel valued and respected are more likely to have a positive impression of the company.
2. Trust and Credibility: Building trust with candidates helps establish credibility. When candidates trust the recruiter, they're more likely to open up about their needs, concerns, and expectations.
3. Better Matches: Relationship building allows recruiters to understand the candidate's goals, values, and motivations. This leads to better matches between candidates and companies, resulting in higher job satisfaction and reduced turnover.
4. Candidate Loyalty: Candidates who have a positive experience with a recruiter are more likely to become loyal to the company. This can lead to referrals, repeat business, and a strong employer brand.
5. Competitive Advantage: In a competitive job market, relationship building can be a key differentiator. Companies that prioritize candidate relationships are more likely to attract top talent.

How to Build Relationships with Candidates

1. Communication: Regular, clear, and transparent communication is essential. Keep candidates informed about the status of their applications and provide feedback when possible.

2. Personalization: Tailor your approach to each candidate's needs and preferences. Show genuine interest in their goals, values, and motivations.
3. Active Listening: Listen attentively to what candidates say and respond thoughtfully. This helps build trust and understanding.
4. Empathy: Show empathy and understanding when candidates face challenges or concerns. This helps build a strong connection and establishes the recruiter as a trusted advisor.
5. Follow up: Follow up with candidates after the hiring process to ensure they're satisfied with their new role. This helps build a long-term relationship and encourages referrals.

Benefits for Recruiters

1. Increased Candidate Engagement: Relationship-building leads to increased candidate engagement, resulting in a more invested and interested candidate pool.
2. Improved Candidate Quality: By understanding the needs and goals of candidates, recruiters can identify top talent and make more effective matches.
3. Reduced Time to Hire: Relationship building can streamline the hiring process, reducing time to hire and improving the overall efficiency of the recruitment process.
4. Enhanced Employer Brand: A positive candidate experience can enhance the employer brand, attracting top talent and reducing recruitment costs.

Conclusion

Relationship building with candidates is a crucial aspect of successful recruitment. By prioritizing candidate relationships, recruiters can create a positive experience, foster trust, and make more effective matches. This approach leads to increased candidate engagement, improved candidate quality, and a competitive advantage in the job market. By investing time and effort into relationship building, recruiters can reap long-term benefits and drive business success.

Self Assessment

1. What's one takeaway from the recruiter story that resonated most with you?
2. Successful recruiters consistently used what tools or habits?
3. How did one recruiter turn a challenge into a breakthrough?
4. What is one small system or process you could implement this week?
5. Why is it helpful to learn from other recruiters' journeys?

Reflective Prompt: Think of your journey so far—what's one challenge you've overcome in recruitment that others might learn from? Please write it down and consider sharing it with your network to help someone else grow.

APPENDICES

Boolean Cheat Sheet by Role Type.

Here are some Boolean search strings for various roles:

Technical Roles

- DevOps Engineer: `(DevOps OR SRE OR "site reliability") AND (Python OR Ruby OR Java OR GO OR Node OR Scala OR C OR C++ OR C#) AND (Docker OR Kubernetes OR Ansible OR Chef OR Puppet OR Salt OR Terraform) AND (AWS OR "Amazon Web Services" OR Azure OR GCP OR "Google Cloud Platform")`
- Java Developer: `(developer OR SDE OR engineer OR programmer OR MTS OR "member of technical staff") AND Java AND (Spring OR JSF OR Hibernate OR Struts OR Play OR Grails)`
- Python Developer: `(developer OR SDE OR engineer OR programmer OR MTS OR "member of technical staff") AND Python AND (Django OR Flask)`
- Backend Developer: `(application OR API OR microservices OR "server-side") AND (Python OR Ruby OR Java OR GO OR Node OR Scala OR C OR C++ OR C#) AND (Spring OR Rails OR Django OR Flask)`
- Frontend Developer: `JavaScript AND (React OR Angular OR Vue)`

Quality Assurance

- QA Analyst: `(tester OR QA OR "quality assurance" OR SDET OR "software development engineer in test" OR "test automation" OR "automation tester" OR "automation engineer")`

Data and Analytics

- Data Engineer: `("Data engineer" OR "extract transform and load" OR ETL OR "data pipelines" OR "data ingestion" OR "data processing") AND (Python OR Scala OR Java)`
- Data Warehouse Engineer: `SQL AND (Bash or Python or scripting) AND ("business intelligence" OR SAS OR Tableau) AND ("data warehouse" OR "data warehousing" OR Redshift OR Snowflake OR Oracle)`

- Machine Learning Engineer: `(Python OR Scala OR Java) AND (Scikit-learn OR TensorFlow OR Pytorch OR Keras OR "machine learning" OR ML)`

Other Roles

- Web Developer: `HTML AND CSS AND (JavaScript OR AJAX OR "content management system" OR Drupal OR WordPress)`
- Mobile Developer: `(developer OR engineer OR programmer) AND (mobile OR Android OR iOS OR Objective-C OR Swift OR Cocoa OR Cocoa-Touch OR SwiftUI OR XCode)`
- Sales Engineer: `("sales") AND (engineer OR developer OR programmer) AND (Python OR Java OR Ruby OR Go OR Node OR C#)`

US Visa and Work Authorisation Reference Table.

	Duration	Employer requirements	Limitations
H1B Visa	Up to 3 years, with the option to extend to 6 years.	Must sponsor the employee, file a petition with USCIS, and demonstrate that the employee has specialized knowledge.	Limited to working for the sponsoring employer, the cap on the number of visas issued annually (85,000)
Optional Practical Training (OPT)	12 months (24 months for STEM fields)	No sponsorship is required, but the employer must be enrolled in E-Verify	Limited to students on F-1 visas, must be related to the field of study
Curricular Practical Training (CPT)	Varies, typically semester long	No sponsorship is required.	Limited to students on F-1 visas, this must be an integral part of the curriculum.
Employment Authorization Documents (EAD)	Varies (1-3 years), depending on immigration status.	No sponsorship is required.	Not transferable between employers, eligibility depends on immigration status.
Green card	Permanent residence	Sponsorship is required for most categories (EB-1, EB-2, EB-3)	Annual caps on the number of visas issued and lengthy processing times.

Templates for Candidate Outreach Email, intake forms, and candidate submission summary format (for VMS and direct clients.)

Candidate Outreach Email Template.

Subject: Exciting Opportunity with [Company Name]

Dear [Candidate Name],

I came across your profile and was impressed with your experience In [industry/field]; I have an exciting opportunity with [Company Name] that you would be an excellent fit for.

Job Details:

- Job Title: [Job Title]
- Company: [Company Name]
- Location: [Location]
- Job Description: [Briefly describe the job and key responsibilities]

If you're interested, I'd love to schedule a call to discuss further. Please let me know your availability.

Best,
[Your Name]

Intake Form Template.

Candidate Intake Form

Section 1: Candidate Information

- Name: _____
- Contact Information: _____
- Current Job Title: _____
- Relevant Experience: _____

Section 2: Job Requirements

- Job Title: _____
- Industry: _____

- Key Skills: _____
- Location: _____

Section 3: Additional Information

- Any specific requirements or preferences?
- Availability for interviews or start date?

Candidate Submission Summary Format

Candidate Information:

- Name: [Candidate Name]
- Contact Information: [Candidate Contact Information]

Job Details:

- Job Title: [Job Title]
- Company: [Company Name]
- Location: [Location]

Summary of Qualifications:

- [Briefly describe the candidate's relevant experience and qualifications]

Key Strengths:

- [List key strengths and skills that align with the job requirements]

Next Steps:

- [Outline next steps, such as interviews or additional screening]

For VMS Clients:

- Vendor Management System (VMS) ID: [VMS ID]
- Client Contact Information: [Client Contact Information]

For Direct Clients:

- Client Contact Information: [Client Contact Information]
- Job Order Number: [Job Order Number]

These templates can be customized to fit your specific recruitment needs and branding requirements.

Time zone Chart.

Conversion chart for major US time zones, with sample working hour overlaps for Asia, Africa, Europe and Oceania.

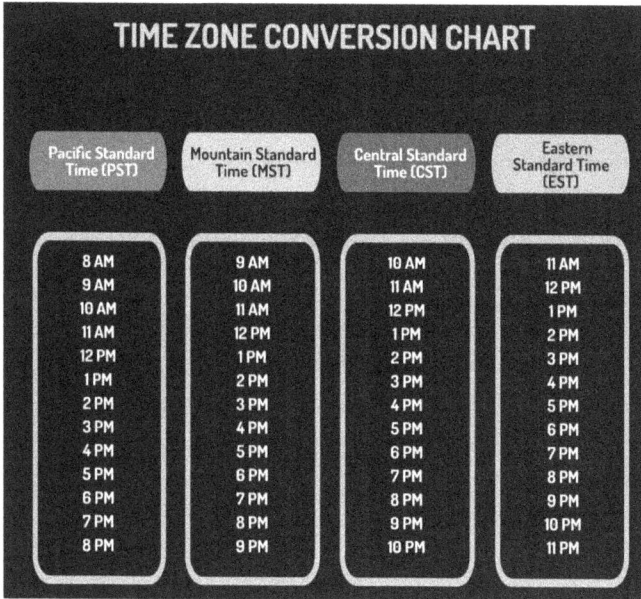

TIME ZONE CONVERSION CHART

Pacific Standard Time (PST)	Mountain Standard Time (MST)	Central Standard Time (CST)	Eastern Standard Time (EST)
8 AM	9 AM	10 AM	11 AM
9 AM	10 AM	11 AM	12 PM
10 AM	11 AM	12 PM	1 PM
11 AM	12 PM	1 PM	2 PM
12 PM	1 PM	2 PM	3 PM
1 PM	2 PM	3 PM	4 PM
2 PM	3 PM	4 PM	5 PM
3 PM	4 PM	5 PM	6 PM
4 PM	5 PM	6 PM	7 PM
5 PM	6 PM	7 PM	8 PM
6 PM	7 PM	8 PM	9 PM
7 PM	8 PM	9 PM	10 PM
8 PM	9 PM	10 PM	11 PM

Recommended tools and communities.

Here are some recommended tools and communities for recruiters:

ATS/CRM/VMS Tools

- Applicant Tracking Systems (ATS):
- Workday
- BambooHR
- Greenhouse
- Lever
- iCIMS
- Customer Relationship Management (CRM) Tools:
- Salesforce
- HubSpot

- Zoho CRM
- Vendor Management Systems (VMS):
- Beeline
- WorkMarket
- IQNavigator
- HeroClix

Recruiting Newsletters

- RecruitingDaily: A daily newsletter covering recruitment news and trends
- SourceConA newsletter focused on sourcing and recruitment strategies
- Recruitment Grapevine: A newsletter covering recruitment industry news

Online Communities

- SourceCon: A community for sources and recruiters
- RecruitingDaily: A community for recruitment professionals
- SHRM (Society for Human Resource Management): A community for HR professionals
- Reddit's r/Recruitment: A community for recruiters and HR professionals
- LinkedIn Groups: Various groups for recruitment and HR professionals

Free Training Platforms

- Coursera: Offers free online courses on recruitment and HR
- edX: Offers free online courses on HR and recruitment
- LinkedIn Learning (formerly (link unavailable)): Offers video courses on recruitment and HR
- Indeed's Hiring and Management Academy: Offers free training on recruitment and hiring
- HR Bartender: A blog and community offering HR and recruitment advice and training.

Glossary of 100+ industry-specific terms

Here's a list of 100 industry-specific terms for staffing and recruitment:

1. ADA: Americans with Disabilities Act;
2. ADEA: Age Discrimination in Employment Act.
3. Applicant Flow Log: Record of job applicants.
4. Assessment: Evaluation of candidate skills.
5. ATS: Applicant Tracking System
6. Background Check: Verification of candidate's history.
7. Benefits Package: Collection of employee benefits.
8. Bona Fide Occupational qualification (BFOQ): Job requirement justified by business necessity.
9. Boolean: Search logic using operators (AND, OR, NOT)
10. Business Continuity Plan (BCP): Plan ensuring business continuity
11. Business Partner: HR professional working closely with business leaders.
12. Candidate Experience: Candidate's perception of the hiring process
13. Contractor: Non-employee worker
14. CPT: Curricular Practical Training
15. CRM: Customer Relationship Management
16. Cultural fit: Alignment of employee values with company culture.
17. Development: Process enhancing employee skills
18. Direct Hire: Permanent employee hire
19. Diversity Recruitment: Strategies promoting diversity in hiring.
20. Diversity, Equity, and Inclusion: Initiatives promoting diversity and inclusion.
21. EAD: Employment Authorization Document
22. EEO: Equal Employment Opportunity
23. F-1 Visa: Student visa for international students
24. FLSA: Fair Labor Standards Act
25. Green Card: Permanent resident card.
26. H1B Visa: Specialty occupation visa

27. HR: Human Resources
28. I-9 Form: Employment eligibility verification form
29. IC: Independent Contractor
30. IT; Information Technology
31. Job Board: Website for posting job openings
32. KPO: Knowledge Process Outsourcing
33. LMS: Learning Management System
34. Net 30: Payment term (payment due within 30 days)
35. OFCCP; Office of Federal Contract Compliance Programs
36. Onboarding: Process of integrating new employees
37. OPT: Optional Practical Training
38. OSHA; Occupational Safety and Health Administration
39. PO: Purchase Order
40. Recruiter: Professional responsible for hiring
41. RFP; Request for proposal
42. SOW: Statement of work
43. Sourcing: Identifying potential candidates
44. Talent Acquisition: Process of finding and hiring talent
45. Temp: Temporary employee
46. VMS: Vendor management system
47. W-4 Form; Employees withholding certificate.
48. Wage Compression: Narrowing wage gaps between employees
49. Work Authorization: Documents authorizing work in the US
50. Affirmative action: Policies promoting equal opportunity
51. Employee handbook; Guide outlining company policies
52. Employment eligibility verification: Process verifying work authorization
53. Equal Pay Act: Law requiring equal pay for equal work
54. Exit interview: Interview with departing employee
55. FMLA: Family and Medical Leave Act
56. Gig economy: Economy based on freelance and contract work
57. Headhunter: Recruiter specialized in high-level positions
58. HRIS: Human resource information system
59. Immigration Reform and Control Act (IRCA): Law regulating employment eligibility
60. Job description: Document outlining job responsibilities
61. Key performance indicator (KPI): Metric measuring performance

62. Labor law: Regulations governing employment
63. Layoff; Temporary or permanent job termination
64. Managed service provider (MSP): Company managing recruitment process
65. New hire orientation: Process of introducing new employees to the company
66. Offer letter: Document extending job offer
67. On-the-job training: Training provided at the workplace
68. Outplacement: Services supporting terminated employees
69. Performance Management: Process evaluating employee performance
70. Predictive Analysis: Analysis forecasting future outcomes
71. Probationary Period: Trial period for new employees
72. Recruitment Marketing: Strategies promoting job openings
73. Referral program: Program incentivizing employee referrals
74. Retention; Strategies reducing employee turnover
75. Salary range: Range of pay for a specific job
76. Screening: Process evaluating candidate qualifications
77. Selection: Process choosing candidates for hire
78. Staffing agency: Company providing temporary or permanent staffing
79. Succession planning: Process identifying future leaders
80. Talent management: Process managing employee development
81. Talent pipeline: Pool of potential candidates
82. Termination: End of employment
83. Time to hire: Metric measuring hiring process efficiency
84. Title VII: Law prohibiting employment discrimination
85. Training and development: Programs enhancing employee skills
86. Turnover rates: Metric measuring the rate of employee turnover
87. User experience (UX): Candidate's experience with the hiring process
88. Vesting period: Period required for employee benefit
89. Workforce planning: Process analyzing workforce needs
90. Work-life Balance: Balance between work and personal life
91. HIPAA: Health insurance portability and accountability act
92. DevOps Engineer: Professional ensuring smooth operation of software systems
93. Bandwidth: Amount of data transmitted over a network
94. API: Application programming interface

95. Data scientist: Professional analyzing and interpreting complex data
96. UX Designer: Professional designing user-centred experiences
97. RFP: Request for proposal
98. Neurodiversity in the workplace: Accommodating neurodiverse employees
99. Talent development programs: Programs enhancing employee skills
100. Workforce planning tools: Software and tools used for workforce planning
101. Virtual onboarding: Online onboarding process
102. 360-degree feedback; Comprehensive evaluation of employee performance.

Clearance and Government Hiring Considerations

Security Clearances

- Public Trust: A level of clearance required for positions involving sensitive information or fiduciary responsibilities.
- Secret: A higher level of clearance is required for positions involving classified information that could cause severe damage to national security if disclosed without authorization.
- Top Secret: The highest level of clearance required for positions involving classified information that could cause exceptionally grave damage to national security if disclosed without authorization.

US Federal Roles Requiring Citizenship

- Federal Law Enforcement: Positions in agencies like the FBI, DEA, and ATF often require US citizenship.
- Intelligence Community: Roles in agencies such as the CIA and NSA typically require US citizenship and security clearances.
- Department of Defense: Many positions within the DoD require US citizenship and security clearances.

Guidelines for Screening Candidates

- Verify Citizenship: Confirm the candidate's US citizenship through documentation, such as a passport or birth certificate.
- Conduct Background Checks: Perform thorough background checks to identify any potential security risks.
- Assess Clearance Eligibility: Determine whether the candidate is eligible for the required security clearance level.
- Evaluate Adjudicative Criteria: Assess the candidate's suitability for a security clearance based on factors like financial history, substance abuse, and foreign contacts.
- Clearance Process: Understand the clearance process and timeline to ensure candidates are processed efficiently and effectively.

- Clearance Levels: Ensure candidates understand the different clearance levels and their corresponding requirements.
- Compliance: Ensure compliance with relevant regulations and guidelines when hiring candidates for roles requiring security clearances.

REFERENCES

Cascio, W. F. (2020). Managing Human Resources: Productivity, quality of work life, profits (10th ed.). McGraw-Hill Education.

Collings, D.G., & Mellahi, K. (2009). Strategic talent management: A review and research agenda. Human Resource Management Review, 19 (4), 304-313

Masselos, J. (2025). The 15 Best Recruiting tools to use. Retrieved from (Toggl.com).

Miscellaneous. (2024). The Impact of Globalization on Human Resource Management. Retrieved from (Skuad.io).

Travis, S. (2023). Navigating Global Labor Relations: Challenges and Strategies in a Multinational Workforce. Retrieved from (commercebulletin.com).

Bonifacio, R. (2024). Strategies for success in managing an international workforce. Retrieved from (shiftbase.com).

Capital, H. (2024). Talent Acquisition technology trends: From generative AI for recruiting to skills-based hiring. Retrieved from (www2.deloitte.com).

Pronix. (2023). The future of talent Acquisition: 5 key trends for 2024 and beyond. Retrieved from (pronixinc.com).

Jay, S. (2025). Types of Compensation: A 2025 guide for HR. Retrieved from (aihr.com).

HR Guide. (n.d). Compensation. Retrieved from (hr-guide.com).

RecruiterFlow. (n.d). 18 Best Candidate Sourcing tools for recruiters in 2025. Retrieved from (recruiterflow.com).

Defilipipo, T. (n.d). Refining search using Boolean operators. Retrieved from (unr.edu).

Rongione, A. (2024). Boolean Operators. Retrieved from (quillbot.com).

Montesa, M. (n.d). Phenom Library. Retrieved from (phenom.com)

RecruiterFlow. (n.d). 15 best AI recruiting tools. Retrieved from (recruiterflow.com).

Final Words of Encouragement

As you close this book, you might be wondering what lies ahead. The truth is, recruitment is more than just a job; it's a calling to connect people with opportunity. You have now seen the industry from every angle, understood the tools and trends, and equipped yourself with both the strategy and the heart required to thrive.

A Career that Creates Careers:

This book wasn't just about placing candidates or meeting quotas; it was about building a meaningful path for yourself while helping others do the same. As a recruiter, you are not just filling roles; you are creating a ripple effect that touches businesses, families, and dreams. Every role you help fill has the potential to change someone's life.

Think about it: when you place a candidate in a role that is a good fit for them, you are not just giving them a job – you are giving them a sense of purpose and fulfilment. You are helping them build a career that aligns with their passions and values. You are supporting their growth and development, empowering them to make a meaningful contribution to their organization.

And it's not just about the candidate, too; it's also about the business. When you help a company find the right talent, you are helping them achieve their goals and objectives. You are supporting their growth and success and enabling them to make a positive impact in their industry.

But the ripple effect goes even further. When a candidate is happy and fulfilled in their role, they are more likely to be a positive influence on their colleagues and community. They are more likely to be a supportive partner, parent, or friend. And when a business is booming, it is more likely to be a positive force in its community, creating jobs and opportunities for others.

So, as a recruiter, there is no way you can say you're just filling roles; you're creating a chain reaction of positivity and impact. You are helping to build a better world, one role at a time. And that is an enriching experience.

By understanding the impact you can have as a recruiter, you can approach your work with a sense of purpose and meaning. You can see the potential for growth and development, not just for your candidates and clients but also for yourself. And you can take pride in knowing that you are making a difference in people's lives.

You Are Now Ready to Enter the Market

You have now built knowledge, refined your tools, and discovered what truly matters in global IT recruitment. Now, it is your time to step forward. Whether you're looking to land your first job in recruitment, grow your influence as a talent advisor, or launch your practice, this is your moment!!

You have spent time learning about the industry, understanding the latest trends and technologies, and developing the skills you need to succeed. You have refined your tools and strategies, and you're ready to put them into action. You are equipped with the knowledge and expertise to navigate the complex world of global IT recruitment, and you're confident in your ability to make a real impact.

So, what's next? It's time to take the leap and pursue your goals. Whether you are looking to break into the industry, advance your career, or start your own business, this is your moment. You have the skills, the knowledge, and the passion, so now it's time to take action.

If you're looking to land your first job in recruitment, you already have a solid foundation to build on. You can highlight your knowledge and skills in your resume and cover letter, and be prepared to discuss your understanding of the industry and your approach to recruitment.

If you are looking to grow your influence as a talent advisor, you've also got a great starting point. You can leverage your expertise to build relationships with clients and candidates and position yourself as a thought leader in the industry.

If you are looking to launch your practice, you've got a strong foundation to build on. You can use your knowledge and skills to develop a successful business strategy and build a reputation as a trusted and effective recruiter.

Whatever your goals, this is your moment to shine. You have the talent, the drive, and the expertise; now it's time to take action and make your mark in the industry.

Believe in Yourself and Your Mission

Believe in the value of connecting people to possibility. Trust that, with consistency, empathy, and curiosity, you can build a fulfilling and impactful career. You have the power to make a difference in people's lives, and that's an enriching experience.

Your Impact Goes Beyond Recruitment

As a recruiter, you are not just matching candidates with job openings; you are helping people find purpose and fulfilment. You're contributing to the growth and success of businesses, and you're making a positive impact on the world.

Stay Curious, Stay Adaptable

The recruitment industry is constantly evolving, and it's essential to stay curious and adaptable. Keep learning, growing, and innovating, and you'll stay ahead of the curve.

You've Got This!

You've got the knowledge, the skills, and the passion. Now, it's time to take action. Go out there and make a difference. Build a career that creates careers, and watch the impact ripple out into the world. This book provides everything you need.

About the Author

Jay Barach is a seasoned expert in talent acquisition, IT operations, and project management, currently serving as the Vice President of IT Operations & Recruitment at Systems Staffing Group, Inc., Pennsylvania. With a multidisciplinary background and a strong foundation in engineering, Jay bridges the gap between technology and human resources, bringing strategic clarity to complex hiring landscapes.

Jay has authored multiple research papers published by IEEE, ACM, and Springer, and has contributed book chapters to both Springer and Bloomsbury publishing houses. His writing and speaking engagements span international conferences on topics such as AI, cybersecurity, HR technology, and workforce development.

He holds certifications in Diversity, Equity & Inclusion (DEI), Project Management (PMP), and is a Certified Sales Engineer (CSE). Jay also has an extensive technical background, having earned multiple Cisco certifications, including passing qualifying exams for CCIE in Routing & Switching, Security, Data Center, and CCDE.

Jay is an active member of IEEE (Senior Member), PMI, ACM, SHRM and the North American Association of Sales Engineers, and also serves as a judge for several global innovation and HR technology awards.

This is his first book dedicated to mentoring aspiring talent acquisition professionals and global recruiters serving the U.S. industry.

www.ingramcontent.com/pod-product-compliance
Lightning Source LLC
Chambersburg PA
CBHW071541210326
41597CB00019B/3077